DOGS

DOMESTICATION HISTORY, BEHAVIOR AND COMMON HEALTH PROBLEMS

ANIMAL SCIENCE, ISSUES AND PROFESSION

Additional books in this series can be found on Nova's website under the Series tab.

Additional e-books in this series can be found on Nova's website under the e-book tab.

ANIMAL SCIENCE, ISSUES AND PROFESSIONS

DOGS

DOMESTICATION HISTORY, BEHAVIOR AND COMMON HEALTH PROBLEMS

KATHERINE M. COHEN

AND

LUCAS R. DIAZ

EDITORS

New York

NOTICE TO THE READER

LIBRARY OF CONGRESS CATALOGING-IN-PUBLICATION DATA

Dogs : domestication history, behavior and common health problems / editors, Katherine M. Cohen and Lucas R. Diaz.
  pages cm
 Includes index.
 ISBN: 78-1-62808-530-3 (soft cover)
 1. Dogs. 2. Dogs--Diseases. I. Cohen, Katherine M. II. Diaz, Lucas R.
 SF426.D636 2013
 636.7--dc23
                        2013026857

Published by Nova Science Publishers, Inc. † *New York*

CONTENTS

PREFACE

In this book the authors discuss the domestication history, behavior and common health problems of dogs. Topics include the epidemiological trends and neuropathological manifestations of canine distemper virus; behavior and welfare of dogs; marketing and the domestication of dogs; responsible ownership and behaviors; advances in the canine coproparasitological examination; and an examination of ovariohysterectomy (spaying) and healthy longevity in dogs.

Chapter 1 – Canine distemper virus (CDV) is Morbillivirus that produces different clinicopathological syndromes in dogs and other domestic animals. Canine distemper (CD) has been described in all continents, and is being controlled in most developed countries due to vaccination. However, CD is still a serious problem in developing countries where it is the principal cause of mortality in urban canine populations. This manuscript initially describes the epidemiological aspects of CD, the dissemination of this infectious agent into non-conventional host, and its impact on wildlife and the domestic canine population. The latter part of the manuscript is an overview of common and atypical clinicopathological manifestations of CDV in dogs.

Chapter 2 – The domestic dog (*Canis familiaris*) as we know nowadays is the only specie, among the thirty-eight belonging to the family Canidae, that had been fully domesticated and is in our companionship for more time, when compared with our other favorite pet, the cat. Fossil evidences showed that since the Paleolithic age dogs are linked with human. This mutualistic relation allowed the selection of the ancestral *Canis lupus* (wolf) to occur artificially, somehow, because of this proximity and in favor of some characteristics according to the use for man (hunting, farm animals guard, companionship, etc.). Despite the great variety of races, that diverge in morphological aspects, the healthy, biological and psychological dog needs are general for all, emphasizing one or other particularity on each race. Pet dogs usually live in restrict environments with their owners, who must ensure that their physical and psychological conditions are satisfactory. For that it is essential the concerning for the welfare of the animal through the knowledge of the behavior of the species. With knowledge of dog behavior and with the right tools to provide better welfare for these animals (environmental enrichment techniques), various diseases can be avoided.

Chapter 3 – If we think in Marketing with ethics and professionalism, we can see it as an effective tool in the continuous pursuit of quality in the provision of services in the veterinary segment. What we see today is a common prejudice on the part of professionals in this area, whose concern is limited to caring for animals, especially dogs. We cannot lose sight of the business, the company that was formed to make a profit, obviously the care of these animals is of paramount importance, but the company needs to survive and grow so that new services are offered to satisfy those consumers and customers. The survival of these companies is really possible, with profit and ethics, focusing on dog's health and welfare.

Chapter 4 – The responsible ownership has been increasingly broached in the modern society, mainly due to the growth of emotional bond between owners and their pets. The lack of responsible ownership can lead to many problems for society, including the overgrowth of animal population, contributing to the increase of stray animals and consequently greater risk of zoonosis. In addition, the lack of responsible ownership can lead to an emergence of animal behavioral disorders due to the owner's carelessness. Among these disorders behavioral changes such as aggression, depression, decreased socialization with other animals and self-mutilation habits can be observed. In this context, the main issue allied to possible solutions will be broached for the subject.

Chapter 5 – In this review of advances in the parasitological examination of feces of humans and animals, are three important projects: two techniques that provide diagnostic improvements, and a project that stands out for its originality in the world. The first project is the *TF-Test* technique (Three Fecal Test), which is able to identify a large numbers of parasitic structures in fecal material from humans and animals (dogs and sheep). This technique showed high diagnostic efficacy, surpassing by more than 24% of sensitivity the association of conventional techniques and a commercial kit. The second project concerns the new *TF-Test Modified* technique, which, according to partial research results, demonstrates higher sensitivity than the *TF-Test* technique and other standard routine laboratory techniques. Finally, the third featured project is the automation of diagnosis of intestinal parasites by computerized image analysis. Preliminary studies have shown that this diagnostic system has high accuracy in terms of sensitivity, specificity, efficiency and concordance index κ, in the automatic detection of the 15 most prevalent parasitic species in Brazil. The results presented here demonstrate the diagnostic efficiency of *TF-Test* and *TF-Test Modified* parasitological techniques, and the future perspective of an automated diagnostic system unprecedented in the world.

Chapter 6 – Currently, we notice an increasingly close contact between humans and animals. A significant portion of the world population demonstrates affective needs or choose to live alone and prefer to have a pet as company, such as a dog, which plays an important role in these conditions, by its ease domestication. With this approach, nowadays is extremely important that the Veterinary Clinics and Pet Shops offer to their customers, i.e., the owners of these animals, effective conditions, by training courses and responsible ownership, among others, that make possible the raise, coexistence, that is good for both, owner and pet. The Management dawns as a master tool in organizing, planning and controlling, in order that Clinics and Enterprises, in general, look at this more ethically and offer good business, as well as to its customers and consumers.

In: Dogs   ISBN: 978-1-62808-530-3
Editors: K. M. Cohen and L. R. Diaz   © 2013 Nova Science Publishers, Inc.

Chapter 1

CANINE DISTEMPER VIRUS: EPIDEMIOLOGICAL TRENDS AND NEUROPATHOLOGICAL MANIFESTATIONS

*Selwyn Arlington Headley, Alice Fernandes Alfieri
and Amauri Alcindes Alfieri*
Laboratory of Animal Virology, Department of Veterinary Preventive
Medicine, Universidade Estadual de Londrina, Rodovia Celso Garcia Cid,
Londrina, PR, Brazil

ABSTRACT

Canine distemper virus (CDV) is Morbillivirus that produces different clinicopathological syndromes in dogs and other domestic animals. Canine distemper (CD) has been described in all continents, and is being controlled in most developed countries due to vaccination. However, CD is still a serious problem in developing countries where it is the principal cause of mortality in urban canine populations. This manuscript initially describes the epidemiological aspects of CD, the dissemination of this infectious agent into non-conventional host, and its impact on wildlife and the domestic canine population. The latter part of the manuscript is an overview of common and atypical clinicopathological manifestations of CDV in dogs.

HISTORICAL PERSPECTIVE

The first description of canine distemper (CD) occurred in South America [1], because CD was probably known as a highly contagious disease of dogs since 1746 in Peru [1, 2], and was observed by Spanish scientists visiting South America in 1748 [3]. Additionally, historical, epidemiological, paleopathological and molecular findings suggest that CD originated from measles virus [3].

After the distemper epizootics in South America, CD was probably exported to Europe [1, 3], and then disseminated across the European continent beginning with Spain and England in 1760, Italy and Ireland in 1764, and then Russia in 1770 [1]. In 1763, 900 dogs reportedly died of CD in Madrid during a single day [1]. A large-scale canine mortality associated with CD also occurred in Louisiana, USA, but it remained obscure if the dissemination occurred via contamination of dogs originating from Europe or South America [1].

In 1870, CD was described as the scourge of the canine race in the United Kingdom since dogs of all breeds and ages were affected; healthier animals were more severely infected, mongrels were less affected, and corneal ulceration was described [4]. In 1890, 90% of affected dogs from several British municipalities died due to CD, with clinical manifestations of anorexia, fever, dryness of the nostrils, purulent oronasal discharge, diarrhea, purulent bronchopneumonia associated with a bacterial agent observed microscopically [5].

The virus-induced etiology of CD was proposed by the French veterinarian Henri Carré in 1905; but the first classical description of the clinical signs was attributed to Edward Jenner in 1908 [6, 7]. However, the viral induced theory was contested by several early investigators who proposed a bacterium, *Haemophilus bronchisepticus*, as the etiologic agent, since this pathogen was observed within the respiratory system of healthy dogs and those necropsied with manifestations of CD during outbreaks of canine distemper virus (CDV) infection [8]. The viral theory was not accepted until the results of the excellent experimental studies of 1926 by Dunkin and Laidlaw realized in ferrets were known in which six strains of CDV were described [2, 8].

Dunkin and Laidlaw studied CD thoroughly in ferrets; the disease in ferrets was transmissible to dogs, and vice versa, and they concluded that both animals were infected by the same pathogen [8]. CD as described by Dunkin and Laidlaw was an acute, infectious, febrile disease with an incubation period of four days, and was characterized by coryza, conjunctivitis, a diphasic elevation in body temperature, severe gastrointestinal alterations, discrete respiratory symptoms, but with infrequent observations of neurological manifestations of disease [8].

However, between 1940 and 1950 the problem of plurality of the virus arose, caused predominantly by the emergence of the encephalitic syndrome, which was not completely described by Laidlaw and Dunkin [2]. A new distemper syndrome, termed hard pad disease, was described by Rubarth in 1947, which differed from conventional CD due to the development of hyperkeratosis of the footpads and the frequent manifestations of encephalitis [8].

ETIOLOGY

Canine distemper virus (CDV) genus *Morbillivirus*, family *Paramyxoviridae* [7], is highly contagious and is transmitted to susceptible hosts predominantly by aerosols. CDV is antigenically related to Rinderpest virus, phocine distemper virus, equine morbillivirus, cetacean morbillivirus, and Pest des petit ruminants virus [9]. CDV is a pleomorphic 150-250 nm diameter virion, with a single-stranded negative-sense RNA that is enclosed in a helically symmetrical nucelocapsid [7, 10]. The virion proteins of CDV include three nucleocapsid proteins: an RNA-binding protein (N) previously referred to as nucleocapsid protein (NP), a phosphoprotein (P), and a polymerase protein (L); and three membrane proteins: a matrix protein (M); one fusion protein (F), and an attachment hemagglutinin (H) protein [7, 11].

The H protein is fundamental for infection, since it recognizes compatible ligands on the surface of host cells [7], and activates the F protein by tissues-specific proteases resulting in infection [12]. There is approximately a 10% variation in the amino acid sequences of the H protein, resulting in biological effects on the virus-host reactions [13]. Further, the H protein is one of the most variable morbillivirus proteins [14], and has been extensively used to evaluate phylogenetic relationships of CDV lineages worldwide [12-15]. While an adequate immune response by the host to the H protein might prevent infection [16].

EPIDEMIOLOGICAL ASPECTS OF CANINE DISTEMPER INDUCED-INFECTIONS

Susceptible Animal Populations

Although CD occurs worldwide, infection within canine populations of developed countries is significantly reduced due to vaccination [17], but the disease continues to be one of the main causes of canine mortality in underdeveloped countries [18, 19]. CDV-induced infections have crossed the original species barrier with disease occurring in members of the *Canidae, Procyonidae, Mustelidae, Hyaenidae, Ursidae, Viverridae, and Felidae* families [18, 20-22]. A list of animals know to have been naturally and experimentally infected by CDV is given in Table 1.

The mammalian host that can serve as reservoir for CDV and consequently disseminate the virus in susceptible mammalian hosts seems to be different in distinct geographical locations. In the USA, CDV is endemic in raccoons (*Procyon lotor*), who are considered as the most likely host to serve as a reservoir for this virus by which susceptible dogs and wildlife can become infected due to their interaction with raccoons in urban and semirural areas [13]. Phylogenetic evidences have suggested that the raccoons epidemics in the USA were caused by two genetically distant American CDV lineages [23]. Further, raccoons were incriminated as the source of infection in an outbreak of CDV in California, USA that resulted in the death of 17 large felids [24]. CDV was suspected as responsible for the decline in the population of the island fox (*Urocyon littoralis catalinae*) within the Santa Catalina Island, California [25]. In Europe, martens (*Martes* spp.) are considered as possible intermediate wildlife reservoirs of CDV [26, 27], and were attributed as sources of infection that resulted in the death of 19 large felids in Switzerland [26]. Moreover, recent studies have described the occurrence of unusual mortality of wildlife in several European countries due to CDV [28]. Additionally, the introduction of CDV-infected masked palm civets (*Paguma larvala*) into a National Park was considered as responsible for the mortality of raccoon dogs (*Nyctereutes procyonoides*) in Japan [29].

Another mammalian host that has been repeatedly associated with the dissemination of CDV into susceptible animal populations is free-ranging or mongrel dogs. Mongrels live in close proximity with humans and are likely to become contaminated by CDV-infected dogs [27]. Free-ranging or hunting dogs were considered as responsible for the dissemination of CDV within the declining population of the endangered European mink (*Mustela lutreola*) in southwestern France [27]. The scavenging habit of rural dogs or mongrels facilitates the ingestion of contaminated remains of dead raccoons left aimlessly along roadways and already infected with CDV [13], thereby serving as a vector to disseminate the disease to canine hosts that were not adequately vaccinated.

There are few descriptions of CDV-infection resulting with wildlife mortality in Brazil [18], and probably South America, suggesting that wildlife species might not be the reservoir of CDV within these geographical locations. The first descriptions of CDV-induced mortality in autochthonous Brazilian wildlife occurred in manned wolves (*Chrysocyon brachyurus*), greater Grisons (*Galictis vittata*), and the crab-eating fox (*Cerdocyon thoin*) that were maintained at a Zoological Park in São Paulo [30]. Since that initial report, there have not been documented descriptions of CDV-induced disease in Brazilian wildlife until the recent confirmation of CDV affecting members of the *Canidae* and *Felidae* families [31-34]. Canids affected including the crab-eating, *Cerdocyon thous* [32, 33], hoary, *Lycalopex vetulus* [31], and the pampas fox, *Pseudalopex gymnocercus* [33]. Felids with CDV-induced infection was restricted to one jaguar, *Panthera leo*, and a puma, *Puma concolor* [34]. Alternatively, seroepidemiological surveys realized in Serra do Cipó National Park, Minas Gerais [35], and the Amazon city of Salvaterra, Marajó Island, Pará did not detect the presence of CDV in wild canids [36]. Nevertheless, 66% (46/70) of dogs native to the Serra do Cipó National Park [35], and 9% (2/23) of their counterparts residing in Salvaterra [36], demonstrated seroreactivity to circulating CDV antibodies. More recently, CDV antibodies (13/14) were detected in wild maned wolves (*Chrysocyon brachyurus*) Galheiro Natural Private Reserve, Minas Gerais, southeastern Brazil [37].

CDV has also been associated with disease in wildlife from other South American countries, including the crab-eating fox (*Cerdocyon thous*) from the El Palmar National Park, Argentina, that probably became infected due to the contact with unvaccinated mongrels from adjacent residents [38]. Additionally, neutralizing CDV antibodies were detected in Geoffroy's cats from two additional Parks in Argentina [39]. In Chile, the domestic dogs is considered as the source of infection for wildlife populations maintained within the Fray Jorge National Park, located at the Coquimbo regions [40].

Consequently, the epidemiological situation in South America might be completely different from that of North America or Europe, whereby there is no need for a wildlife species to disseminate CDV between infected and susceptible animal populations [18]. Nevertheless, the importance of CDV in urban and rural populations cannot be underscored. Further, CD has been classified as class 3 Emerging Infectious Disease within continental USA and Africa, since CDV endangers a wide host of carnivores, might be related to the extinction of the African wild dog and populations of black-footed ferrets, and threatens the survival of the Ethiopian wolf [41]. In Europe, CDV might be threatening the existence of the European mink, *Mustela lutreola* [27], and the Iberian lynx, *Lynx pardinus* [42]. The exact effects of CDV on wildlife populations in South America still have to be determined.

Table 1. Animals susceptible to infection by canine distemper virus[1]

Family Affected	Type of Infection	
	Natural	Experimental
Canidae	Dog, raccoon dog, dingo, foxes (bat-eared, kit, red, and grey), coyote, jackal, wolf, South American bush-dog	Dog, ferret
Mustelidae	Weasel, ferret, polecat, mink, skunk, badger, marten, otter, stoat, wolverine, grison	Mink
Felidae	Lion, leopard, bobcat, margay, cheetah, tiger, puma, jaguar, lynx, domestic cat, spotted hyena, and Geoffroy's cat	Domestic cat
Viverridae	Masked palm civet, binturong, linsang, fossa, and genet	
Ailuridae	Lesser panda	
Suidae	Pig, peccary	Pig
Procyonidae	Coati, kinkajou, racoon, bassariscus	
Muridae		Mouse, rat
Cricetidae		Hamster
Ursidae	Bears (polar, grizzly, brown, and black), giant panda	
Herpestidae	Mongoose, meerkat	
Miscellaneous	Nonhuman primate, man?	Nonhuman primate

[1]Adapted from: Summers, B.A., Appel, M.J.G. Aspects of canine distemper virus and measles virus encephalomyelitis. *Neuropath. Appl. Neurobiol.*, *20*, 525–534, (1994). Osterhaus et al., Morbillivirus infections of aquatic mammals: newly identified members of the genus. *Vet. Microbiol. 44*, 219-227, (1995).

Phylogenetic Relationships

Phylogenetic analyses based on the H gene have demonstrated that there are distinct lineages of CDV circulating worldwide; the phylogenetic relationship of these linages is shown in Figure 1. These include the following linages: Asia-1, Asia-2, America-1, America-2, Artic-like, European wildlife,

Europe [12], Africa [15], and Alpine wildlife [43]. Additionally, distinct lineages of CDV were identified circulating within Argentina [44], Brazil [17], and Uruguay [45], and collectively forms the South American clade [17, 45]. Moreover, a new genotype of CDV has been reported circulating in Mexico [19].

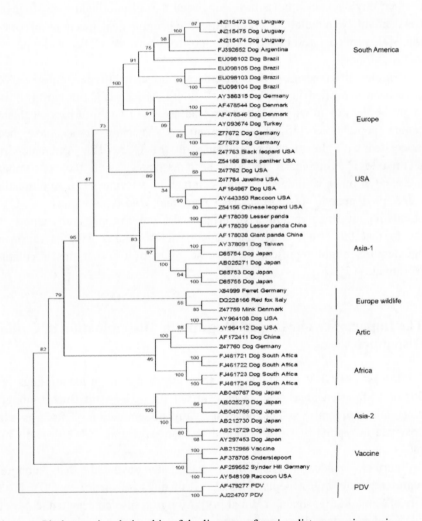

Figure 1. Phylogenetic relationship of the lineages of canine distemper virus using selected sequences of the H protein gene. The tree was constructed with the MEGA 5.10 software, using the neighbor-joining method with 1,000 bootstraps replicates. The GenBank accession numbers, the animal/strain, and the country of origin are given. Phocine distemper virus (PDV) was used as the out-group.

Recent studies have demonstrated that the Brazilian cluster of CDV strains have an amino acid variation between 1.7 and 4% with their European counterparts [17].

Additionally, the combined South American and European lineages of CDV might represent the largest strains circulating worldwide [17, 46]. Further, phylogenetic studies have suggested that distinct wild-type strains of CDV might be circulating within different geographical locations of Brazil [46, 47], probably resulting in antigenic drift between CDV wild-type and vaccine strains.

Similar differences between vaccine and novel wild-type strains of CDV have been identified in the USA [23], South Africa [15], Europe [48], and might be the key to explain the several outbreaks of CD in urban vaccinated canine populations [15, 17]. Alternatively, vaccine strains have been associated with the raccoons epidemics in the USA [15]. Antigenic drift relative to CD outbreaks has also been associated with the concomitant presence of other viral agents within commercial vaccines [15]. Additionally, these phylogenetic results suggest that the wild-type strains of CDV circulating in Brazil might have originated from Europe, and support the theory that the South American clade of CDV strains with its European counterparts might represent the largest group of CDV strains circulating worldwide [17].

The Importance of Mongrel Dogs in the Dissemination of Canine Distemper Virus

In cases of CDV infections described in urban cities affecting dogs [49, 50] or in forested areas with disease in wildlife [31-34] in Brazil, mongrel dogs are frequently associated with viral dissemination. This was also observed in reports of wildlife mortality from Argentina [38, 39], and Chile [40].

Further, epidemiological data have demonstrated that mongrels represented the largest populations (54%) of dogs infected with CDV in the city of Santa Maria, southern Brazil [49]. Mongrels also contributed to 44% of CDV-induced cases diagnosed in Porto Alegre, another city of southern Brazil [51], and 30% of all dogs with CDV from Belém, Pará northern Brazil [50]. Elevated indices of CDV infections in mongrels are directly related to their lifestyle. This is because of the following: mongrels are more likely to roam the streets in semi-urban and rural districts, have unknown vaccination status, and are more susceptible to infections than their purebred counterparts. Additionally, these dogs are more apt to have sporadic contacts with wildlife species and susceptible dogs in rural areas, have scavenging habits that are comparable to raccoons, and might be the likely candidate to serve as reservoirs of CDV in South America. Further, purebred dogs are more rigorously vaccinated against CDV than are their mongrel counterparts, and epidemiological data suggested that purebred dogs were not as frequently infected by CDV relative to mongrels [49]. Although purebred dogs are infected by distemper, dolichocephalic breeds seem more susceptible to CDV than their brachycephalic counterparts do [49, 52]. Moreover, mixed-breed and mongrel dogs were considered as being at risk of infection by CDV in Argentina [53].

In Finland, where the incidence of distemper is relatively low with no reported cases since the 1994-1995 outbreak [54], the most recent cases of CDV infection were attributed to contact between susceptible pups and roaming dogs from Russia [55] and Estonia [56]. In the latter case, the dog also had concomitant Tyzzer's disease and intestinal coccidiosis [56]. Additionally, the roaming habit of mongrels was associated with infestation by *Dioctophyma renale* in dogs from southern Brazil [57], concomitant infections of CDV, *Toxoplasma gondii,* and ehrlichiosis [58], and dual infections of CDV and *T. gondii* [59] in dogs from the city of Botucatu, southwest Brazil. Similar reports have been described with CDV-induced secondary *Bordetella bronchiseptica* bronchopneumonia in puppies from southern Brazil [60], and *Nocardia asteroides* infection from Sudan [61]. Further, a seroepidemiological survey realized in Pelotas, southern Brazil, demonstrated that 68% of dogs positive for CDV had street access and 58% of all positive dogs were mongrels [62]. Recently, a 43-days-old puppy was coinfecetd with CDV, canine parvovirus-2, canine adenovirus A 1 and 2, as well as *T. gondii* [46], and probably contracted these infectious agents either via the infected mother or in contact with unvaccinated infected dogs soon after birth.

The spill-over effect as occurred in the USA [41], was attributed to the dissemination of canine parvovirus, canine adenovirus, and canine coronavirus infections between infected mongrels and susceptible wild canids native from a National Park in southwest [35], and to CDV-induced disease in large felids from State Parks in Midwest Brazil [34]. This phenomenon was also responsible for the dissemination of CDV between roaming dogs and wildlife maintained in parks within Argentina [38, 39] and Chile [40]. Therefore, the large number of mongrels within rural and semi-urban cities of South America would serve not only as reservoirs and distributors of CDV, but for other infectious disease agents, thereby excluding the necessity of having a wildlife species serving as an intermediate host for CDV in these geographical locations, as is described in North America [13], and Europe [26, 27]. Free roaming domestic dogs were also incriminated as being responsible for the spill-over effect of CDV to black-backed jackals (*Canis mesomelas*) in Namibia, Africa [63], to the Serengeti wildlife epidemic in Tanzania [64], and possibly to the Iberian lynxes, whose population is estimated to be approximately only 200 individuals [42]. Moreover, recent studies done in South Korea have demonstrated that free-ranging dogs or foxes were responsible for the dissemination of CDV in a population of raccoon dogs [65]. Similar results have emerged from Taiwan and Vietnam, where mongrels were associated with the dissemination of CDV to domestic and wild felids [66].

Epidemiological data suggest that in Brazil, and probably other South American countries, CDV might be equally endemic in dogs from different age groups. This then raises the obvious question as to why dogs, although protected by attenuated vaccine, frequently demonstrate elevated indices of infection by CDV. Although vaccination failures could be partially associated with this phenomenon, it was proposed that asymptomatic mongrels infected with a wild-type strain of CDV might be disseminating distemper within urban canine populations of Brazil [18]. Further, dogs without the typical clinical manifestations of distemper have been diagnosed with CDV by RT-PCR within several Brazilian cities [67-69]. Restriction fragment length polymorphism analyses of commercial vaccines used in Brazil and wild-type strains of the H gene of CDV in Brazilian cities have suggested that there are molecular differences existing between the commercial vaccine strains of CDV circulating in southern Brazil and those strains attributed to clinical disease in susceptible dogs [47].

This might have resulted in antigenic drift between strains of CDV circulating within Brazil. Moreover, 22% of all dogs vaccinated in south-western Brazil were infected by CDV [70]. Additionally, infected dogs that have been protected by attenuated CDV vaccines are probably contaminated due to exposure of wild-type CDV [13]. Alternatively, it was suggested that the large amount of CD observed in adult dogs might be related to inadequate vaccination [71]; nevertheless, these animals would theoretically have some form of protection and would have to become in contact with an infected animal to develop the disease. These arguments further stressed the need for molecular epidemiological studies be designed in all South American countries to further identify and discriminate the strains of CDV that are circulating in urban canine populations relative to those used in commercial vaccine production, in an attempt to elucidate the cause of the elevated mortality in dogs protected with attenuated vaccination protocols. Additionally, the possibility of manufacturing new vaccines must also be considered as novel strains of CDV are being discovered [15].

Seasonal Occurrence

Studies realized in southern Brazil have suggested that the highest prevalence of dogs diagnosed with CDV occurred during the winter-spring period, with comparatively reduced prevalence during summer-autumn [49, 72]. Epidemiological data obtained from northern Brazil demonstrated that CD was more prevalent during the cold, humid, rainy season [50]. Moreover, seasonal variations of CD with elevated indices during the colder months have also been described in the USA [6, 52], Argentina [53], and India [73]. Although the exact influence of seasonal variations on the occurrence of distemper is obscure, the colder weather facilities the maintenance [6, 49], and increases the survival time of CDV [10], and might result in stress-induced immunosuppression in neonates and recently weaned dogs [49], facilitating infection in naïve animals.

In CDV infection, immunosuppression probably occurs due to the lymphotrophism achieved by the binding of the H viral protein to cells that express the human equivalent to CD 150, signaling lymphocytic activation molecule (SLAM) [7, 74]. Although SLAM is expressed in healthy dogs, it is up regulated in the lymphoid cells of dogs infected with CDV [74], and expressed by thymocytes, activated macrophages, and dendritic cells [7]. Therefore, destruction of cells that express SLAM might be the mechanism associated with immunosuppression in CD. Dogs that are immunocompromised due to CDV infections are apt to develop secondary bacterial [56, 58, 60, 61] or protozoan infections [46, 56, 58, 59], or simultaneous viral infections [46, 75, 76].

CDV has been reported as surviving near freezing temperatures between 0-4 °C for several weeks [10] and environmental stress has been associated with infection [7], which might favor elevated infectious rates during the colder climate. Alternatively, seasonal predominance of distemper in raccoons has been attributed to the mating season [77].

Neuropathological Manifestations of Canine Distemper Virus

Histopathological Diagnosis of Canine Distemper Encephalitis

Although the histopathological lesions associated with canine distemper encephalitis (CDE) are unique, a diagnosis of distemper encephalitis must be based exclusively on the histological patterns associated with characteristic intranuclear/intracytoplasmic inclusion bodies [78]. In CDE, viral inclusion bodies are more frequently observed in acute relative to chronic manifestations of disease, particularly due to the extremely complex, and not fully elucidated, immunopathological reactions associated with chronic distemper encephalitis [74, 79, 80]; readers are encouraged to examine these manuscripts for in-depth reviews of CDV-induced pathogenesis and immunological reactions. In basic terms, the initial lesions of CDE are considered viral mediated, while progression to sclerosing plaque-like alterations observed in chronic CDE is governed by immunopathological reactions [74]. Inclusion bodies are numerous during 10-14 days post infection (PI), and the number of inclusions is rapidly reduced 4-5 weeks PI [78], these periods correspond, respectively, to the histological manifestations of acute and chronic CDE. Although CDV inclusions bodies were described as readily observed in 70 cases of CDE [71], the finding of inclusion bodies is an arduous, painstaking, and time-consuming task particularly in chronic CDE, and as such is highly susceptible to induce false negatives.

The progression of CDE as characterized by histopathology should be based on the previously described patterns [81, 82], which offers a simple and efficient characterization of histopathological alterations that can be readily observed and realized, thereby ensuring repeatability of results. Briefly, it characterizes CDE into three distinct progressive lesions: acute, subacute, and chronic. Acute CDE is characterized by demyelination with discrete to moderate astrogliosis but without perivascular cuffs or inflammatory cells (Figure 2A). Subacute CDE contains moderate demyelination with an influx of inflammatory cells, perivascular cuffing (formed by two or three layers of mononuclear cells), moderate astrogliosis and astrocytosis with discrete accumulations of macrophages (Figure 2B). Chronic CDE consists of demyelination with extensive perivascular cuffings (four or more layers of mononuclear cells), moderate to severe astrocytosis and astrogliosis, marked influx of lymphoplasmacytic and histiocytic inflammatory cells with multinucleated astrocytes (Figure 2C). However, all patterns must demonstrate characteristic eosinophilic inclusion bodies (Figure 2D).

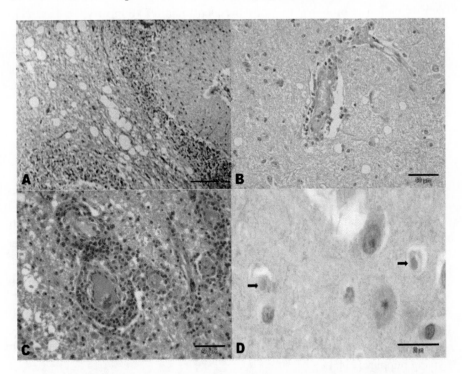

Figure 2. Histopathological features of canine distemper encephalitis. Observe severe white matter demyelination without perivascular cuffing and influx of inflammatory cells in acute distemper encephalitis (A). There is perivascular cuffing formed by at least two layers of mononuclear cells in subacute distemper encephalitis with discrete inflammatory influx (B). Chronic distemper encephalitis is characterized by perivascular cuffs formed by more than four layers of mononuclear cells with severe influx of inflammatory cells (C). Observe characteristic eosinophilic intranuclear inclusion bodies (arrows) within astrocytes (D). Hematoxylin and Eosin stain; Bar: A and C, 100 μm; B, 50 μm; D, 20 μm. (Previously published: Headley, et al. Semina: Ciências Agrárias 33, 1945-1978, 2012).

It must be highlighted that this histological classification should be used only to evaluate the progression of CDE, but is not synonymous for the clinical syndromes associated with CDV. Most syndromes associated with CDE are characterized principally by the neuroanatomical localization of the lesion, the histological features, the neuropathological manifestations, and to some extent, the age of the affected animal. Therefore, histological classification is fundamental to characterize CDV-related syndromes. Nevertheless, irrespective of the histological progression of disease, demyelination is probably the most frequent manifestation of CDE [51, 71]. Histopathological manifestations of CD demyelinating encephalitis are more frequently observed within the cerebellum [71, 81], followed by the diencephalon, frontal lobe of the telencephalon, pons, and the mesencephalon [71]. Additionally, demyelination in CDE is more severe within the white matter of the cerebellum, rostral medullary velum, optic tracts, spinal cord, and around the fourth ventricle, and might be associated with viral dissemination via the cerebrospinal fluid [78].

CDV produces severe clinical manifestations in dogs and other susceptible hosts, resulting in disease to most organ systems, such as the respiratory, cutaneous, urinary, gastrointestinal, and the central nervous system (CNS). The manifestations of CDE are directly related to the strain of virus, the age and the immune status of the affected animal, and the neuroanatomical location affected [7, 20]. These vary with the neuroanatomical location of the lesion, the age and the immunological status of the affected animal, the histological pattern of disease, and viral strain [83-85]. These neurological manifestations of CDV include CDE in immature dogs, multifocal distemper encephalomyelitis in mature dogs, old dog encephalitis, post-vaccinal distemper encephalitis [78, 85-87], polioencephalomalacia [88, 89], chronic relapsing encephalomyelitis [90], and atypical necrotizing encephalitis of young puppies [91].

Canine Distemper Encephalitis in Immature Dogs

This is the most common manifestation of CDV-induced infections [86], and is frequently associated with systemic CD in dogs with clinical manifestations and histopathological demonstrations of disease within, but not restricted to, the CNS, gastrointestinal, respiratory, urinary, and cutaneous systems. In some cases, enamel hypoplasia of developing teeth [92] due to the direct action of CDV on ameloblasts [93], nasal and footpad hyperkeratosis [94], cutaneous abdominal pustules, and myocardial necrosis [92] might be observed. Cutaneous pustules are manifestations of systemic CDV that have been complicated with staphylococcal and streptococcal infections [78]. The neuropathological manifestations of CDE in immature dogs are more frequently observed within the white matter of the cerebellar peduncles, the optic nerve, and the spinal cord [78, 86]. The histopathological lesions are characterized by varying degrees of demyelination (Figure 3A), perivascular cuffings, degeneration of nerve fibers with the formation of spheroids, edema, neuronal necrosis [55, 86], with neuroparenchymal necrosis and infiltration of macrophages in severe cases [86].

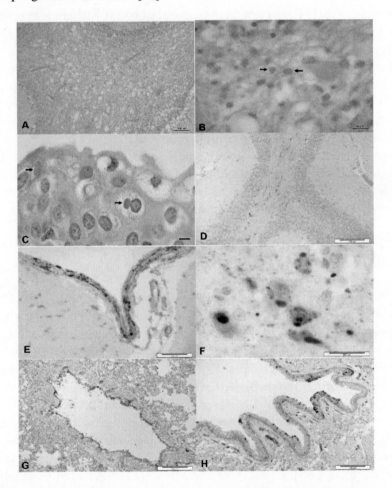

Figure 3. Histopathological and immunohistochemical features of canine distemper encephalitis in immature dogs. Observe white matter demyelination (A) with intranuclear inclusion bodies within astrocytes (B) of the cerebellum and intracytoplasmic inclusion bodies within the transitional epithelium of the urinary bladder (C). They is positive immunoreactivity to antigen of canine distemper virus within the white matter of the cerebellum (D), meninges (E) neurons of the cerebrum (F), and epithelial cells of the lung (G) and urinary bladder (H). A-C, Hematoxylin and eosin stain; D-H, immunoperoxidase. Bar, A, 100 μm; B-C, F, 10 μm; E-H, 200 μm.

In some cases, there are severe accumulations of lymphoplasmacytic inflammatory cells within the choroid plexus and the spinal cord. Intranuclear and/or intracytoplasmic eosinophilic inclusion bodies are readily observed within astrocytes (Figure 3B), neurons, and within glial, ependymal and meningeal cells [55], but are more frequently observed within astrocytes [6, 78]. Intracytoplasmic inclusion bodies are also present in extra-neural tissues such as the transitional epithelium of the urinary bladder (Figure 3C), lung, and renal pelvis. Positive immunoreactivity to CDV antigens within the brain are predominantly and easily observed by IHC within astrocytes of the cerebellar white matter (Figure 3D), neurons of the spinal cord, and even in tissues where inclusion bodies are not easily identified by routine histopathology, such as in the choroid plexus and the meninges (Figure 3E). However, different from the cerebellum, positive immunoreactivity to CDV antigens within the cerebrum seems predominant within neurons relative to astrocytes (Figure 3F). Positive immunoreactivity is also easily demonstrated in non-neural tissues such as the urinary bladder and lung (Figure 3G-H).

Multifocal Distemper Encephalomyelitis in Mature Dogs

This an unusual syndrome of low incidence associated with CDV [67, 86] that occurs in dogs that are between 4-8 years of age [86, 87]. Clinical manifestation of disease is slow and progressive [87], and includes nystagmus, positional ventrolateral strabismus, and spastic tetraparesis [67], but without seizures and personality changes [87]. The histopathological manifestations of disease are restricted to the CNS, being more prevalent within the cerebellum and spinal cord white matter, with sparing of the cerebral cortex [83, 87]. This anatomic predominance of histopathological lesions within the cerebellum and spinal cord with the absence of cerebral involvement differentiates this syndrome from old dog encephalitis, in which the cerebrum is frequently affected [83, 87, 95]. Histological alterations are characterized by multifocal necrotizing to demyelinating nonsuppurative encephalitis [67, 87], associated with rare inclusion bodies [86].

Old Dog Encephalitis (ODE)

ODE is a very rare and unusual manifestation of CDV-induced encephalitis that occurs in dogs that are more than six years of age [86, 87,

95], but dogs as young as one year of age have been diagnosed with this disease [87]. This rare manifestation of CDE was first described by Cordy in 1942 [96], with subsequent reports in the 1970s [97-99] and 80s [83]. The absence of reported cases of ODE during decades after the initial descriptions had resulted in speculation as to the existence of this unique manifestation of CDV since spontaneous cases have not been observed in several institutions [86]. Further, it was suggested that some of the previously reported cases of ODE might have been in fact CDE in older dogs [20]. This uncertainty was probably supported due to the difficulty and successive frustrated attempts to isolate and transmit the virus *in vitro* [96-99], because in ODE, CDV appears to persist in a replication-defective state [20, 87]. However, the disease was experimentally reproduced in a gnotobiotic dog [100], and more recently, a spontaneous case was described in southern Brazil [95]. Consequently, these recent descriptions confirm that ODE exists, but is indeed rarely diagnosed.

Clinically, ODE is characterized by progressive cortical and subcortical dysfunctions with manifestations of mental depression, unresponsive behavior, head pressing [86], circling, swaying, and weaving gait [87], motor incoordination, behavioral changes, and seizures [95]. Systemic manifestations of CDV are absent in ODE [86, 101], and visual impairment might be the initial manifestation of this syndrome [86]. The neuropathological alterations are restricted to the forebrain with marked sparing of the hindbrain (cerebellum and brain stem) and occipital cortex [87, 95]. In the case described in southern Brazil, gross manifestation of disease included loss of distinct demarcation between the white and grey matter, with an irregular surface and marked grey-brown discoloration to the affected tissue; these changes are more easily appreciated when the affected region is compared with the normal tissue (Figure 4A-C).

In ODE, the salient histopathological alterations are restricted to the forebrain and consist predominantly of extensive perivascular cuffings (Figure 5A), severe influx of lymphoplasmacytic inflammatory cells, marked glial proliferation, neuronal necrosis, rarefaction of neuroparenchymal tissue, and accumulations of multinucleated giant cells [83, 95-99]. Most perivascular cuffs contained more than six layers of mononuclear cells; multinucleated giant cells frequently contain inclusion bodies (Figure 5B-C). Inclusion bodies in ODE are considered either to be frequent [83, 95, 98, 99], infrequent [96], or absent [100]. Immunohistochemical studies of the case described in southern Brazil readily demonstrated CDV+ antigens predominantly within neurons relative to astrocytes (Figure 5D), but with weak immunoreactivity within syncytial cells [95]. Additionally, the multinucleated giant cells previously described in other cases of ODE [98, 100], were strongly immunoreactive to vimentin and macrophage antigen (Figure 5E-F), but were negatively labelled with glial fibrillary acidic protein, CD3, and CD79a, suggesting that these cells are more likely of a monocytic origin [95]. Additionally, the N gene (formerly NP) of CDV was successfully extracted from formalin-fixed paraffin-embedded tissues, amplified by RT-PCR, the amplicons sequenced, and the obtained nucleotide sequences have been deposited in GenBank [95], representing the first characterization of this gene in ODE.

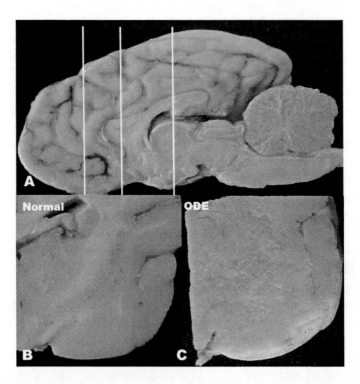

Figure 4. Gross findings in a case of old dog encephalitis; formalin-fixed sections. The normal brain is shown with the neuroanatomical locations (1, 2, and 3) where gross manifestations were predominant (A). The normal (B) was taken section for comparison at transverse location 3 and compared with the similar but affected region (C). Observe the marked absence of differentiation between the white and grey matter due to inflammation and degeneration (C). (Previously published: Headley, et al. Semina: Ciências Agrárias 33, 1945-1978, 2012).

Post-Vaccinal Distemper Encephalitis

This form of CDE has been associated with neurological manifestations observed between 8-20 days after the routine administration of specific strains of CDV vaccines in dogs [102, 103] and wild animals [104, 105].

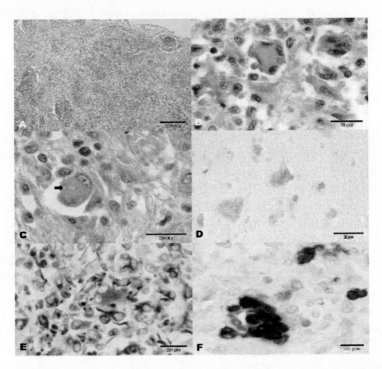

Figure 5. Histopathological and immunohistochemical findings in old dog encephalitis. There is severe destruction of brain tissue due to extensive perivascular cuffs and severe influx of inflammatory cells (A). Several multinucleated giant cells were observed admixed within the inflammatory infiltrate (B) and some of these (arrow) contained viral inclusion bodies (C). Observe positive immunoreactivity to antigens of canine distemper virus within neurons and astrocyte (D). Multinucleated giant cells demonstrated positive immunoreactivity to vimentin (E) and macrophage antigen (F). A-C, Hematoxylin and eosin stain; D-F, immunoperoxidase. Bar, A, 200 μm; B-E, 20 μm; F, 10 μm. (Previously published: Headley, et al. Semina: Ciências Agrárias 33, 1945-1978, 2012).

The clinical manifestations associated with this syndrome include diarrhea, anorexia, aggressive behavior, ocular discharge, prostration, posterior incoordination later, circling, blindness, and seizures terminating in death within 2-3 days after vaccination [102, 103, 105]. Although the neuropathological alterations in post-vaccination encephalitis are disseminated, the lesions are pronounced at the brainstem [102, 105], the cerebellar peduncles [105], cerebral cortex, thalamus, and spinal cord [103]. The associated histopathological alterations are multifocal, being constituted by malacia, spheroids, neuronal degeneration, neuronophagia, gliosis, and perivascular cuffings with intracytoplasmic and/or intranuclear inclusion bodies [102, 103, 105].

CDV-Associated Polioencephalomalacia

The pathogenesis of polioencephalomalacia (PEM), also known as cortical cerebral necrosis, in dogs is not well-elucidated [85], but has been associated with manifestations of convulsive CDE [88, 89], thiamine deficiency [106], and lead poisoning [107]. Nevertheless, in cases of CDV-induced PEM the affected dog can clinically demonstrate depression, myoclonus, lateral head tilt, circling, blindness [88], muscle tremors, and convulsions [89]. The histopathological alterations associated with CDV-induced PEM are degenerative, symmetrical [89] or bilateral [86], and are observed predominantly at the cerebral cortex [86, 88, 89], with focal lesions at the cerebellum [89]. The histopathological alterations are characterized by hypertrophy and proliferation of endothelial cells and adventitia of vessels, small perivascular cuffings, and neuronal degeneration associated with acidophilic intranuclear inclusion bodies within degenerated neurons and astrocytes at the cerebral cortex [88, 89]. In CDV-induced PEM, demyelinating encephalitis has been observed [88] or is absent [86]. Nevertheless, demyelinating encephalitis in CDV-induced PEM seems to be restricted to the cerebellum [88, 89], and absent in the cerebrum and spinal cord [86].

Chronic Relapsing Encephalomyelitis

This unique neurological manifestation of CDV was described in an 18-month-old dog that presented chronic progressive multiphasic neurological disease [90]. Clinically, this syndrome was characterized by initial manifestations of paresis, increased tendon reflexes, deficient conscious proprioception in the right pelvic limb and mild paresis in the left pelvic limb.

After a period of two weeks, the disease progressed to bilateral paretic and ataxic of pelvic limbs, altered conscious proprioception in the pelvic limbs, with complete paraplegia and absence of voluntary control of urination and defecation [90]. Three distinct neuropathological lesions were observed by histopathology: cystic necrosis of the spinal cord, chronic demyelinating encephalitis the cerebellum, and acute demyelinating encephalitis at the pons [90].

Atypical Necrotizing Distemper Encephalitis of Young Puppies

This is a recently diagnosed atypical manifestation of systemic CD [91]. The disease occurred in a litter of puppies that were less than one month of age, being characterized by clinical manifestations of bilateral forebrain dysfunction (decreased to absent postural reactions, normal spinal reflex, seizures, and decreased level of consciousness without cranial nerves deficits). Neuropathological manifestations of disease were observed only in the gray and white matters of the forebrain (telencephalon and diencephalon) with sparing of the hindbrain (cerebellum, cerebellar peduncles, pons, and medulla oblongata). Histopathological alterations were restricted to the forebrain, and were characterized by asymmetrical necrosis with severe influx of monocytic and histiocytic inflammatory cells, syncytial formation, severe astrocytic response associated with intranuclear and intracytoplasmic eosinophilic inclusion bodies within glial cells. Further, intralesional antigens of CDV were demonstrated by IHC, and the N gene of CDV was amplified by RT-PCR from fresh tissue fragments from the brain and visceral organs. Atypical necrotizing encephalitis of young pups is unique in that it affects only the forebrain of dogs with systemic CD; hence, the neuroanatomical localization of this disease differentiates this syndrome from other known manifestations of CDE. However, the pathological manifestations of this syndrome must be differentiated primarily from those associated with canid herpesvirus 1 (CaHV-1)-induced infections in neonates. CaHV-1 produces multifocal necrosis that is frequently observed within several parenchymal organs (liver, kidney, lungs, thymus, and spleen) as well as the intestines and brain [87, 108], with discrete manifestations of disease affecting the stomach, pancreas, omentum, retina, and myocardium [108]. The neuropathological manifestations of CaHV-1 encephalitis are characterized by severe nonsuppurative meningoencephalitis primarily affecting the cerebellum and brainstem [87], with cerebral cortical necrosis and necrosis of the thalamus, and hemorrhage, lymphoplasmacytic inflammatory infiltrate and perivascular cuffing within the cerebrum, cerebellum, thalamus, pons and medulla [108].

REFERENCES

[1]   Blancou, J. Dog distemper: imported into Europe from South America? *Hist Med Vet*, 29, 35-41 (2004).

[2]   Howell, DG. Immunization of the dog. *Can Vet J*, 6, 127-136 (1965).

[3]   Uhl, EW; Kelderhouse, C; Blick, J; Hogan, RJ. Evidence of a New World origin for canine distemper *The FASEB Journal*, 26, 631.634 (2011).

[4]   McBridge, On ulcers of the cornea in dogs suffering from distemper. *Br Med J*, 159 (1870).

[5]   Millais, E. The pathogenic microbe of distemper in dogs, and its use for protective Inoculation. *Br Med J*, 1, 856-860 (1890).

[6]   Appel, MJ. Canine distemper virus, in *Virus infections of vertebrates*, Appel, M.J., Editor, Elsevier: Amsterdam. 133-159 (1987).

[7]   MacLachlan, NJ; Dubovi, EJ. *Paramyxoviridae*. 4th ed. Fenner's veterinary virology. San Diego, California: Academic Press. 299-325 (2011).

[8]   Gledhill, AW. Some veterinary diseases of medical interest. *Br Med Bul*, 9, 237-241 (1953).

[9]   Quinn, PJ; Markey, BP; Carter, ME; Donnelly, MJ; Lonard, FC. *Paramyxoviridae. Veterinary microbiology and microbial disease.* Ames, Iowa: Blackwell Sciences Ltd (2004).

[10] Greene, CE; Vandevelde, M. Canine distemper, in *Infectious diseases of the dog and cat.*, Greene, C.E., Editor, Elsevier: St Louis. 25-42 (2012).

[11] Lamb, RA; Kolakofsky, D. Paramyxoviridae: the viruses and their replication, in *Fields Virolog*, Fields, D.M., Knipe, D.M., and Howley, P.M., Editors, Lippincott-Williams & Wilkins,: Phialdephia. 1305-1443. (2001).

[12] McCarthy, AJ; Shaw, MA; Goodman, SJ. Pathogen evolution and disease emergence in carnivores. *Proceedings.The Royal Society Biological Sciences,* 274, 3165-3174 (2007).

[13] Kapil, S; Allison, RW; Johnston, 3rd, L; Murray, BL; Holland, S; Meinkoth, J; Johnson, B. Canine distemper virus strains circulating among North American dogs. *Clin Vaccine Immunol*, 15, 707-712 (2008).

[14] Pardo, ID; Johnson, GC; Kleiboeker, SB. Phylogenetic characterization of canine distemper viruses detected in naturally infected dogs in North America. *J Clin Microbiol*, 43, 5009-5017 (2005).

[15] Woma, TY; van Vuuren, M. Bosman, AM; Quan, M; Oosthuizen, M. Phylogenetic analysis of the haemagglutinin gene of current wild-type canine distemper viruses from South Africa: lineage Africa. *Vet Microbiol*, 143, 126-132 (2010).

[16] Martella, V; Elia, G; Buonavoglia, C. Canine distemper virus. *Vet Clin North Am. Small Anim Pract*, 38, 787-797, vii-viii (2008).

[17] Negrão, FJ; Gardinali, NR; Headley, SA; Alfieri, AA; Fernandez, MA; Alfieri, AF. Phylogenetic analyses of the hemagglutinin gene of the southern Brazil wild-type strains of canine distemper virus. *Genet Mol Res*, (2013).

[18] Headley, SA; Amude, AM; Alfieri, AF; Bracarense, APRFL; Alfieri, AA. Epidemiological features and the neuropathological manifestations of canine distemper virus-induced infections in Brazil: a review. *Semina Ciências Agrárias*, 33, 1945-1978 (2012).

[19] Gamiz, C; Martella, V; Ulloa, R; Fajardo, R; Quijano-Hernandez, I; Martinez, S. Identification of a new genotype of canine distemper virus circulating in America. *Vet Res Commun*, 35, 381-390 (2011).

[20] Summers, BA; Appel, MJ. Aspects of canine distemper virus and measles virus encephalomyelitis. *Neuropathol Appl Neurobiol*, 20, 525-534 (1994).

[21] Osterhaus, AD; de Swart, RL; Vos, HW; Ross, PS; Kenter, MJ; Barrett, T. Morbillivirus infections of aquatic mammals: newly identified members of the genus. *Vet Microbiol*, 44, 219-227 (1995).

[22] Deem, SL; Spelman, LH; Yates, RA; Montali, RJ. Canine distemper in terrestrial carnivores: a review. *Journal of Zoo and Wildlife Medicine*, 31, 441-451 (2000).

[23] Lednicky, J; Dubach, J; Kinsel, M; Meehan, T; Bocchetta, M; Hungerford, L; Sarich, N; Witecki, K; Braid, M; Pedrak, C; Houde, C. Genetically distant American canine distemper virus lineages have recently caused epizootics with somewhat different characteristics in raccoons living around a large suburban zoo in the USA. *Virology Journal*, 1, 2 (2004).

[24] Appel, MJ; Yates, RA; Foley, GL; Bernstein, JJ; Santinelli, S; Spelman, LH; Miller, LD; Arp, LH; Anderson, M; Barr, M; et al., Canine distemper epizootic in lions, tigers, and leopards in North America. *J Vet Diagn Invest*, 6, 277-288 (1994).

[25] Timm, SF; Munson, L; Summers, BA; Terio, KA; Dubovi, EJ; Rupprecht, CE; Kapil, S; Garcelon, DK. A suspected canine distemper epidemic as the cause of a catastrophic decline in Santa Catalina Island foxes (*Urocyon littoralis catalinae*). *J Wildl Dis*, 45, 333-343 (2009).

[26] Myers, DL; Zurbriggen, A; Lutz, H; Pospischil, A. Distemper: not a new disease in lions and tigers. *Clin Diagn Lab Immunol*, 4, 180-184 (1997).

[27] Philippa, J; Fournier-Chambrillon, C; Fournier, P; Schaftenaar, W; van de Bildt, M; van Herweijnen, R; Kuiken, T; Liabeuf, M; Ditcharry, S; Joubert, L; Begnier, M; Osterhaus, A. Serologic survey for selected viral pathogens in free-ranging endangered European mink (*Mustela lutreola*) and other mustelids from south-western France. *J Wildl Dis*, 44, 791-801 (2008).

[28] Origgi, FC; Plattet, P; Sattler, U; Robert, N; Casaubon, J; Mavrot, F; Pewsner, M; Wu, N; Giovannini, S; Oevermann, A; Stoffel, MH; Gaschen, V; Segner, H; Ryser-Degiorgis, MP. Emergence of canine distemper virus strains with modified molecular signature and enhanced neuronal tropism leading to high mortality in wild carnivores. *Vet Path*, 49, 913-929 (2012).

[29] Machida, N; Kiryu, K; Oh-ishi, K; Kanda, E; Izumisawa, N; Nakamura, T. Pathology and epidemiology of canine distemper in raccoon dogs (*Nyctereutes procyonoides*). *J Comp Pathol*, 108, 383-392 (1993).

[30] Rego, AAMS; Matushima, ER; Pinto, CM; Biasia, I. Distemper in Brazilian wild canidae and mustelidae: case report. *Braz J Vet Res Anim Sci*, 34, 256-258 (1997).

[31] Megid, J; Teixeira, CR; Amorin, RL; Cortez, A; Heinemann, MB; de Paula Antunes, JM; da Costa, LF; Fornazari, F; Cipriano, JR; Cremasco, A; Richtzenhain, LJ. First identification of canine distemper virus in hoary fox (Lycalopex vetulus): pathologic aspects and virus phylogeny. *J Wildl Dis*, 46, 303-305 (2010).

[32] Megid, J; de Souza, VA; Teixeira, CR; Cortez, A; Amorin, RL; Heinemman, MB; Cagnini, DQ; Richtzenhain, LJ. Canine distemper virus in a crab-eating fox (Cerdocyon thous) in Brazil: case report and phylogenetic analyses. *J Wildl Dis*, 45, 527-530 (2009).

[33] Hübner, SdO; Pappen, FG; Ruas, JL; Vargas, GDÁ; Fischer, G; Vidor, T. Exposure of pampas fox (*Pseudalopex gymnocercus*) and crab-eating fox (*Cerdocyon thous*) from the Southern region of Brazil to Canine distemper virus (CDV), Canine parvovirus (CPV) and Canine coronavirus (CCoV). *Braz Arch Biol Technol*, 53, 593-597 (2010).

[34] Nava, AF; Cullen, Jr. L; Sana, DA; Nardi, MS; Filho, JD; Lima, TF; Abreu, KC; Ferreira, F. First evidence of canine distemper in Brazilian free-ranging felids. *EcoHealth*, 5, 513-518 (2008).

[35] Curi, NHdA; Araújo, AS; Campos, FS; Lobato, ZIP; Gennari, SM; Marvulo, MFV; Silva, JCR; Talamoni, SA. Wild canids, domestic dogs and their pathogens in Southeast Brazil: disease threats for canid conservation. *Biodives Conserv*, 19, 3513-3524 (2010).

[36] Courtenay, O; Quinnell, RJ; Chalmers, WS. Contact rates between wild and domestic canids: no evidence of parvovirus or canine distemper virus in crab-eating foxes. *Vet Microbiol*, 81, 9-19 (2001).

[37] de Almeida Curi, NH; Coelho, CM; de Campos Cordeiro Malta, M; Magni, EM; Sabato, MA; Araujo, AS; Lobato, ZI; Santos, JL; Santos, HA; Ragozo, AA; de Souza, SL. Pathogens of wild maned wolves (*Chrysocyon brachyurus*) in Brazil. *J Wildl Dis*, 48, 1052-1056 (2012).

[38] Ferreyra, H; Calderon, MG; Marticorena, D; Marull, C; Leonardo, BC. Canine distemper infection in crab-eating fox (*Cerdocyon thous)* from Argentina. *J Wildl Dis*, 45, 1158-1162 (2009).

[39] Uhart, MM; Rago, MV; Marull, CA; Ferreyra Hdel, V; Pereira, JA. Exposure to selected Pathogens in to selected pathogens in Geoffroy's cats and domestic carnivores from central Argentina. *J Wildl Dis*, 48, 899-909 (2012).

[40] Acosta-Jamett, G; Chalmers, WS; Cunningham, AA; Cleaveland, S; Handel, IG; Bronsvoort, BM. Urban domestic dog populations as a source of canine distemper virus for wild carnivores in the Coquimbo region of Chile. *Vet Microbiol*, 152, 247-257 (2011).

[41] Daszak, P; Cunningham, AA; Hyatt, AD. Emerging infectious diseases of wildlife--threats to biodiversity and human health. *Science*, 287, 443-449 (2000).

[42] Meli, ML; Simmler, P; Cattori, V; Martinez, F; Vargas, A; Palomares, F; Lopez-Bao, JV; Simon, MA; Lopez, G; Leon-Vizcaino, L; Hofmann-Lehmann, R; Lutz, H. Importance of canine distemper virus (CDV) infection in free-ranging Iberian lynxes (*Lynx pardinus*). *Vet Microbiol*, 146, 132-137 (2010).

[43] Monne, I; Fusaro, A; Valastro, V; Citterio, C; Dalla Pozza, M; Obber, F; Trevisiol, K; Cova, M; De Benedictis, P; Bregoli, M; Capua, I; Cattoli, G. A distinct CDV genotype causing a major epidemic in Alpine wildlife. *Vet Microbiol*, 150, 63-69 (2011).

[44] Calderon, MG; Remorini, P; Periolo, O; Iglesias, M; Mattion, N; La Torre, J. Detection by RT-PCR and genetic characterization of canine

distemper virus from vaccinated and non-vaccinated dogs in Argentina. *Vet Microbiol*, 125, 341-349 (2007).

[45] Panzera, Y; Calderon, MG; Sarute, N; Guasco, S; Cardeillac, A; Bonilla, B; Hernandez, M; Francia, L; Bedo, G; La Torre, J; Perez, R. Evidence of two co-circulating genetic lineages of canine distemper virus in South America. *Virus research*, 163, 401-404 (2012).

[46] Headley, SA; Alfieri, AA; Fritzen, JTT; Garcia, JL; Weissenböck, H; Silva, AP; Bodnar, L; Okano, W; Alfieri, AF. Concomitant canine distemper, infectious canine hepatitis, canine parvoviral enteritis, canine infectious tracheobronchitis, and toxoplasmosis in a puppy. *J Vet Diagn Invest*, 25, 112 - 118 (2013).

[47] Negrão, FJ; Wosiacki, SH; Alfieri, AA; Alfieri, AF. Restriction pattern of a hemagglutinin gene amplified by RT-PCR from vaccine strains and wild-type canine distemper virus. *Arq Bras Med Vet Zoot*, 58, 1099-1106 (2006).

[48] Martella, V; Cirone, F; Elia, G; Lorusso, E; Decaro, N; Campolo, M; Desario, C; Lucente, MS; Bellacicco, AL; Blixenkrone-Moller, M, Carmichael, LE; Buonavoglia, C. Heterogeneity within the hemagglutinin genes of canine distemper virus (CDV) strains detected in Italy. *Vet Microbiol*, 116, 301-309 (2006).

[49] Headley, SA; Graça, DL. Canine distemper: epidemiological findings of 250 cases. *Braz J Vet Res Anim Sci*, 37, 136-140 (2000).

[50] Guedes, IB; Lima, AS; Espinheiro, RF; Manssour, MB; Cruz, IP; Dias, HLT. Occurrence and geographical assessment of canine distemper in the city of Belém, Pará-Brazil. in *World Small Animal Veterinary Association*. 2009. São Paulo, Brazil.

[51] Sonne, L; Oliveira, EC; Pescador, CA; Santos, AS; Pavarini, SP; Carissimi, AS; Driemeier, D. Pathologic and immunohistochemistry findings in dogs naturally infected by canine distemper virus. *Pesq Vet Bras*, 29, 143-149 (2009).

[52] Gorham, JR. The epizootiology of distemper. *J Am Vet Med Assoc*, 149, 610-622 (1966).

[53] Perez, AM; Marro, V; Schiaffino, L; Pirles, M; Bin, L; Ward. MP. Risk factors associated with canine Distemper in Casilda, Argentina. in *Proceedings of the 10th International Symposium on Veterinary Epidemiology and Economics*. 2003. Vina del Mar, Chile: SciQuest.

[54] Ek-Kommonen, C; Sihvonen, L; Pekkanen, K; Rikula, U; Nuotio, L. Outbreak off canine distemper in vaccinated dogs in Finland. *Vet Rec*, 141, 380-383 (1997).

[55] Headley, SA; Sukura, A. Naturally occurring systemic canine distemper virus infection in a pup. *Brazilian Journal of Veterinary Pathology*, 2, 95 – 101 (2009).

[56] Headley, SA; Shirota, K; Baba, T; Ikeda, T; Sukura, A. Diagnostic exercise: Tyzzer's disease, distemper, and coccidiosis in a pup. *Vet Path*, 46, 151-154 (2009).

[57] Nakagawa, TL; Bracarense, AP; dos Reis, AC; Yamamura, MH; Headley, SA. Giant kidney worm (*Dioctophyma renale*) infections in dogs from Northern Parana, Brazil. *Vet Parasitol*, 145, 366-370 (2007).

[58] Moretti, LdA; Silva, AVd; Ribeiro, MG; Paes, AC; Langoni, H. *Toxoplasma gondii* genotyping in a dog co-infected with distemper virus and ehrlichiosis rickettsia. *Rev Inst Med Trop São Paulo*, 48, 359-363 (2006).

[59] Moretti, LD; Ueno, TE; Ribeiro, MG; Aguiar, DM; Paes, AC; Pezerico, SB; Silva, AV. Toxoplasmosis in distemper virus infected dogs. *Semina Ciências Agrárias*, 23, 85-91 (2002).

[60] Headley, SA; Graça, DL; Costa, MMd; Vargas, ACd. Canine distemper virus infection with secondary *Bordetella bronchiseptica* pneumonia in dogs. *Cienc Rural*, 29, 741-743 (1999).

[61] Fawi, MT; Tag el Din, MH; el-Sanousi, SM. Canine distemper as a predisposing factor for *Nocardia asteroides* infection in the dog. *Vet Rec*, 88, 326-328 (1971).

[62] Hass, R; Johann, JM; Caetano, CF; Fischer, G; Vargas, GD; Vidor, T; Hübner, SO. Antibodies levels against canine distemper virus and canine parvovirus in vaccinated and unvaccinated dogs. *Arq Bras Med Vet Zoot*, 60, 270-274 (2008).

[63] Gowtage-Sequeira, S, Banyard, AC; Barrett, T; Buczkowski, H; Funk, SM; Cleaveland, S. Epidemiology, pathology, and genetic analysis of a canine distemper epidemic in Namibia. *J Wildl Dis*, 45, 1008-1020 (2009).

[64] Cleaveland, S; Appel, MG; Chalmers, WS; Chillingworth, C; Kaare, M; Dye, C. Serological and demographic evidence for domestic dogs as a source of canine distemper virus infection for Serengeti wildlife. *Vet Microbiol*, 72, 217-227 (2000).

[65] Cha, S-Y; Kim, E-J; Kang, M; Jang, S-H; Lee, H-B; Jang, H-K. Epidemiology of canine distemper virus in wild raccoon dogs (*Nyctereutes procyonoides*) from South Korea. *Comparative Immunology, Microbiology and Infectious Diseases*, 35, 497-504 (2012).

[66] Ikeda, Y; Nakamura, K; Miyazawa, T; Chen, MC; Kuo, TF; Lin, JA; Mikami, T; Kai, C; Takahashi, E. Seroprevalence of canine distemper virus in cats. *Clin Diagn Lab Immunol*, 8, 641-644 (2001).

[67] Amude, AM; Carvalho, GdA; Alfieri, AA; Alfieri, AF. Virus isolation and molecular characterization of canine distemper virus by RT-PCR from a mature dog with multifocal encephalomyelit. *Braz J Microbiol*, 38, 354-356 (2007).

[68] Amude, AM; Alfieri, AA; Alfieri, AF. Antemortem diagnosis of CDV infection by RT-PCR in distemper dogs with neurological deficits without the typical clinical presentation. *Vet Res Commun*, 30, 679-687 (2006).

[69] Del Puerto, HL; Vasconcelos, AC; Moro, L; Alves, F; Braz, GF; Martins, AS. Canine distemper virus detection in asymptomatic and non vaccinated dogs. *Pesq Vet Bras*, 30, 132-138 (2010).

[70] Gouveia, AMG; Magalhães, HH; Ribeiro, AL. Canine distemper: occurrence in vaccinated animals and age-group distribution. *Arq Bras Med Vet Zoot*, 39, 539-545 (1987).

[71] Silva, MC; Fighera, RA; Mazzanti, A; Brum, JS; Pierezan, F; Barros, CSL. Neuropathology of canine distemper: 70 cases (2005-2008). *Pesq Vet Bras*, 29, 643-652 (2009).

[72] Borba, TR; Mannigel, RC; Fraporti, CK; Headley, SA; Saito, TB. Canine distemper: epidemiological data Maringá (1998-2001). *ICCesumar*, 4, 53-56 (2002).

[73] Kadaba, D. An Epidemiological Study of Canine Distemper in Mumbai: Bridging the Gap Between Human and Animal Health. *Epidemiology*, 22, S112-S113 (2011).

[74] Beineke, A; Puff, C; Seehusen, F; Baumgartner, W. Pathogenesis and immunopathology of systemic and nervous canine distemper. *Vet Immunol Immunopathol*, 127, 1-18 (2009).

[75] Chvala, S; Benetka, V; Mostl, K; Zeugswetter, F; Spergser, J; Weissenbock, H. Simultaneous canine distemper virus, canine adenovirus type 2, and *Mycoplasma cynos* infection in a dog with pneumonia. *Vet Path*, 44, 508-512 (2007).

[76] Rodriguez-Tovar, LE; Ramirez-Romero, R; Valdez-Nava, Y; Nevarez-Garza, AM; Zarate-Ramos, JJ; Lopez, A. Combined distemper-adenoviral pneumonia in a dog. *Can Vet J*, 48, 632-634 (2007).

[77] Roscoe, DE. Epizootiology of canine distemper in New Jersey raccoons. *J Wildl Dis*, 29, 390-395 (1993).

[78] Caswell, JL; Williams, KJ. Canine distemper, in *Jubb, Kennedy, and Palmer's Pathology of domestic animals.*, Maxie, M.G., Editor, Saunders/Elsevier: Philadelphia. p. 635-638 (2007).

[79] Vandevelde, M; Zurbriggen, A. The neurobiology of canine distemper virus infection. *Vet Microbiol*, 44, 271-280 (1995).

[80] Vandevelde, M; Zurbriggen, A. Demyelination in canine distemper virus infection: a review. *Acta Neuropathol*, 109, 56-68 (2005).

[81] Vandevelde, M; Fankhauser, R; Kristensen, F; Kristensen, B. Immunoglobulins in demyelinating lesions in canine distemper encephalitis. An immunohistological study. *Acta Neuropathol*, 54, 31-41 (1981).

[82] Headley, SA; Soares, IC; Graça, DL. Glial Fibrillary Acidic Protein (GFAP)-immunoreactive astrocytes in dogs infected with canine distemper virus. *J Comp Pathol*, 125, 90-97 (2001).

[83] Vandevelde, M; Kristensen, B; Braund, KG; Greene, CE; Swango, LJ; Hoerlein, BF. Chronic canine distemper virus encephalitis in mature dogs. *Vet Path*, 17, 17-28 (1980).

[84] Summers, BA; Greisen, HA; Appel, MJ. Canine distemper encephalomyelitis: variation with virus strain. *J Comp Pathol*, 94, 65-75 (1984).

[85] Summers, BA; Cummings, JF; De Lahunta, A. *Veterinary Neuropathology*. St. Louis: Mosby. 527 (1994).

[86] Braund, KG. *Clinical syndromes in veterinary neurology*. 2 ed., St. Louis: Mosby(1994).

[87] Maxie, MG; Youssef, S. Nervous system, in *Jubb, Kennedy & Palmer's Pathology of Domestic Animals* M.G., M., Editor, Saunders/Elsevier: Philadelphia. p. 281-457 (2007).

[88] Finnie, JW; Hooper, PT. Polioencephalomalacia in dogs with distemper encephalitis. *Aust Vet J*, 61, 407-408 (1984).

[89] Lisiak, JA; Vandevelde, M. Polioencephalomalacia associated with canine distemper virus infection. *Vet Path*, 16, 650-660 (1979).

[90] Higgins, RJ; Child, G; Vandevelde, M. Chronic relapsing demyelinating encephalomyelitis associated with persistent spontaneous canine distemper virus infection. *Acta Neuropathol*, 77, 441-444 (1989).

[91] Amude, AM; Headley, SA; Alfieri, AA; Beloni, SN; Alfieri, AF. Atypical necrotizing encephalitis associated with systemic canine distemper virus infection in pups. *J Vet Sci*, 12, 409-412 (2011).

[92] Headley, SA; Saito, TB. Simultaneous canine distemper encephalitis and canine parvovirus infection with distemper-associated cardiac necrosis in a pup. *Cienc Rural*, 33, 1075-1080 (2003).

[93] Dubielzig, RR. The effect of canine distemper virus on the ameloblastic layer of the developing tooth. *Vet Path*, 16, 268-270 (1979).

[94] Koutinas, AF; Baumgartner, W; Tontis, D; Polizopoulou, Z; Saridomichelakis, MN; Lekkas, S. Histopathology and immunohistochemistry of canine distemper virus-induced footpad hyperkeratosis (hard Pad disease) in dogs with natural canine distemper. *Vet Path*, 41, 2-9 (2004).

[95] Headley, SA; Amude, AM; Alfieri, AF; Bracarense, AP; Alfieri, AA; Summers, BA. Molecular detection of canine distemper virus and the immunohistochemical characterization of the neurologic lesions in naturally occurring old dog encephalitis. *J Vet Diagn Invest*, 21, 588-597 (2009).

[96] Cordy, DR. Canine encephalomyelitis. *Cornell Vet*, 32, 11-28 (1942).

[97] Adams, JM; Brown, WJ; Snow, HD; Lincoln, SD; Sears, Jr. AW; Barenfus, M; Holliday, TA; Cremer, NE; Lennette, EH. Old dog encephalitis and demyelinating diseases in man. *Vet Path*, 12, 220-226 (1975).

[98] Lincoln, SD; Gorham, JR; Ott, RL; Hegreberg, GA. Etiologic studies of old dog encephalitis. I. Demonstration of canine distemper viral antigen in the brain in two cases. *Vet Path*, 8, 1-8 (1971).

[99] Lincoln, SD; Gorham, JR; Davis, WC; Ott, RL. Studies of old dog encephalitis. II. Electron microscopic and immunohistologic findings. *Vet Path*, 10, 124-129 (1973).

[100] Axthelm, MK; Krakowka, S. Experimental old dog encephalitis (ODE) in a gnotobiotic dog. *Vet Path*, 35, 527-534 (1998).

[101] Jones, TC; Hunt, RD; King, NW. *Veterinary Pathology*. 6 ed., Baltimore: Williams and Wilkins. 1392 (1997).

[102] Hartley, WJ. A post-vaccinal inclusion body encephalitis in dogs. *Vet Path*, 11, 301-312 (1974).

[103] Cornwell, HJ; Thompson, H; McCandlish, IA; Macartney, L; Nash, AS. Encephalitis in dogs associated with a batch of canine distemper (Rockborn) vaccine. *Vet Rec*, 122, 54-59 (1988).

[104] Halbrooks, RD; Swango, LJ; Schnurrenberger, PR; Mitchell, FE; Hill, EP. Response of gray foxes to modified live-virus canine distemper vaccines. *J Am Vet Med Assoc*, 179, 1170-1174 (1981).

[105] Kazacos, KR; Thacker, HL; Shivaprasad, HL; Burger, PP. Vaccination-induced distemper in kinkajous. *J Am Vet Med Assoc*, 179, 1166-1169 (1981).

[106] Read, DH; Jolly, RD; Alley, MR. Polioencephalomalacia of dogs with thiamine deficiency. *Vet Path*, 14, 103-112 (1977).

[107] Zook, BC. The pathologic anatomy of lead poisoning in dogs. *Vet Path*, 9, 310-327 (1972).

[108] Appel, MJ. Canine herpesvirus, in *Virus infections of vertebrates*, Appel, M.J., Editor, Elsevier: Amsterdam. p. 5-15 (1987).

In: Dogs   ISBN: 978-1-62808-530-3
Editors: K. M. Cohen and L. R. Diaz   © 2013 Nova Science Publishers, Inc.

Chapter 2

DOMESTICATION ASPECTS, BEHAVIOR AND WELFARE OF DOGS

Juliana Damasceno[1] and Rachel Stopatto Righetti[2]
[1]Biologist, MSc degree in the Psychobiology Postgraduate Program,
Postgraduate student at the PhD level in the area of Psychobiology
at the University of São Paulo (USP), São Paulo, Brazil.
[2]Graduated in Portuguese, English Language and Literature
from Moura Lacerda University Centre, São Paulo, Brazil.

The domestic dog (*Canis familiaris*) that we currently know is the species among the thirty-eight belonging to the family Canidae that has been fully domesticated and has been our companion far longer than our other favorite pet, the cat. Fossil evidence shows that since the Paleolithic age, dogs have been linked to humans. This mutual relation caused the selection of the ancestral *Canis lupus* (wolf) to occur artificially because of the wolf's proximity to man and because of some of the wolf's useful characteristics (hunting, guarding farm animals, companionship, etc.). Despite the great variety of breeds that morphologically diverge, the health, biological and psychological needs of dogs are common to all breeds, emphasizing one or another characteristic for each breed. Pet dogs generally live in restricted environments with their owners, who must ensure that their physical and psychological needs are satisfied. Thus the welfare of the animal depends on humans' knowledge of the behavior of the species. An understanding of dog behavior and the right tools can provide better care for these animals (environmental enrichment techniques), and various diseases can be avoided.

INTRODUCTION

The current population of dogs in the United States is approximately 70 million, according to the Animal Veterinary Medical Association (AVMA), and in the United Kingdom 23% of all families have a dog, which is equivalent to 8 million dogs, according to the Pet Food Manufacturers' Association (PFMA). In addition to being one of our most important pets, dogs are also used as study models in various veterinary, odontology, medical, neuroscientific, behavioral and other studies. To ensure these animals' physical and psychological health independent of their environment, it is important to understand their evolution, domestication and typical behavior. With understanding and the right tools, it is easy to provide satisfactory and effective conditions for dogs' health and welfare. This chapter provides a general overview of dogs' history from their origins and their proximity to man until today, highlighting the breadth and necessity of these animals' behaviors and indicating some tools to ensure a more complex and psychologically healthy canine life.

HISTORY AND DOMESTICATION

The *Canidae* family, which includes the dog (*Canis familiaris*), began its divergence from the other carnivorous families approximately 50 million years ago (Vilà and Wayne, 1999). The canids are now represented by thirty-eight species living around the world except for Antarctica and some oceanic islands (Clutton-Brock, 1995). Of the pets we now know, the dog was the first to be domesticated. The domestication of the dog occurred even before humans developed a closer contact with production animals and thousands of years before the domestication of our also beloved pet, the cat. Archaeological evidence shows that the domestication of the dog began during the Glacial Era, when human subsistence still depended on hunting, gathering and foraging (Clutton-Brock, 1995).

Davis and Valla (1978) reported on a human fossil found in Israel dated between 10.000-12.000 years ago that was buried with a Canidae puppy, which could have been a domesticated wolf or even a "dog". The human belonged to the Natufians, an Egyptian culture from the Epipaleolitic period.

The Natufian culture comprised hunters and gatherers about to become farmers. The genre of the buried human could not be identified because of its bent position and because it was on its side although the puppy's age was estimated at approximately four or five months at time of death. The human skeleton had its hand on the puppy's chest, demonstrating a strong connection between them, and there has been a special human-dog relationship since that time. Valla (1990) reports on another fossil registration, also found in Israel and belonging to the same Natufian culture, of a human buried beside two Canidae that were perhaps dogs. These and other archaeological sites that had canidae skeletons next to humans confirm that beyond domestication, a close relation between humans and dogs has existed for thousands of years.

Despite clear evidence that dogs have been among us since long ago, several questions emerge when we think about the dog's ancestors and origins. The most accepted and widespread theory, based on the most evidence, is that the dog comes from the domestication of wolves (*Canis lupus*). Although this theory is the most accepted, there are few details regarding how the first dog appeared or how such a variety of breeds developed. Other less clear and more uncertain theories and studies regarding the origins of dogs have been presented such as theories involving the jackal and the coyote as possible ancestors (Roots, 2007). However, facts to support these theories are even more scarce (Savolainen, 2007) than support of the wolf ancestor theory.

The most accepted theory and the one that has the most genetic and morphologic evidence infers that the modern dog began its differentiation from wolves more than 100 thousand years ago (Vilà et al. 1997). The existence of isoenzymatic alleles, polymorphic microsatellites and mitochondrial DNA sequences, which dogs and wolves have in common,, confirm the closeness between them (Wayne and Vilà, 2001). A series of studies regarding aspects of behavior, vocalization, morphology and molecular biology indicate that the main and perhaps the only ancestor of the dog is the wolf (Clutton-Brock, 1995).

If dogs did indeed descend from wolves, how did domestication occur with such wild animals? Domestication is a principle of evolution, a synonym for change, combining genetic changes from one generation to the next as well as change induced by the environment. Evolution affects animals in three main manners: psychologically, morphologically and behaviorally (Price 1984; Roots, 2007). The process of domestication is strongly allied with and influenced by human actions both culturally and biologically. The cultural aspects are related to the incorporation of animals in society; animals are perceived as property to serve humans. The biological aspects are related to the process of evolution, from an animal's wild origins to a friendly and docile relationship with humans (Clutton-Brock, 1992).

According to the literature, wolf domestication began in the Mesolithic period, when humans were still nomadic hunters (Clutton-Brock, 1995). Those wolves that felt the least afraid of men started to follow and help them, first as guards to scare away a possible enemy, either human or animal, and then actively participating in the hunt (Driscoll et al., 2009), collaborating with humans to search and attack prey.

There are several theories regarding wolves' proximity to man. One theory states that when wolves were killed so that their skins could be used for clothing, their puppies, instead of also being killed, were raised by man, becoming more meek and less aggressive. In this manner, wolves were progressively tamed (Clutton-Brock, 1995). Another theory suggests that when wolves helped hunt, the just-born puppies remained in the caves, creating human proximity and establishing a closer relation. Wolves became progressively closer to man and less aggressive (Roots, 2007). As a result of such proximity, increased food sources, and greater safety, the wolves' reproductive rate increased (Udell et al., 2010. Over time, as humans became more sessile and focused on the land, animals gradually became genetically different from their ancestors, adapting to human activities and modifying their morphology and behavior (Driscoll et al., 2009, Udell et al., 2010).

Driscoll et al. (2009) state that domesticating animals and plants has allowed human to accumulate more resources and enjoy better nutrition, which caused the Neolithic revolution to occur, bringing not only an agricultural economy but the development of "urban" life. Neolithic farmers were the first geneticists, selecting domestic animals and plants for their use and developing such characteristics as a more docile temperament, increased jaw and neck strength, and great speed among other characteristics somehow selected by man that began what we know as artificial selection.

The evolution of domesticated species is directly related to artificial selection whereas natural selection has a secondary place in this process (Driscoll et al., 2009). Artificial selection, as the name suggests, is the opposite of natural selection and is related to human actions, economic requirements, aesthetic desires and cultural requirements (Clutton-Brock, 1992). In artificial selection, crossings occur by human interference and may occur during the pre-zygotic stage (when mates are chosen by man) or post-zygotic stage (when the most suitable progeny reproduce differently) (Driscoll et al., 2009). In his book *Domestication,* Roots (2007) states that genotype modifications, facing artificial selection, interrupt naturally established systems, causing changes in some processes such as the genotype removal of recessive alleles. These and other genetic manipulations may cause a loss of diversity or what we now know as genetic erosion, making individuals more susceptible to diseases. Driscoll et al. (2009) infer that artificial selection may occur in two different intensities: "weak" or "strong". "Weak" artificial selection occurs in the post-zygotic stage, whereas "strong" artificial selection may occur in either the pre- or post- zygotic stage, for example, breeding good dairy cow offspring from good females.

Contrary to artificial selection, which is connected to changes caused by humans, natural selection is related to environmental mechanisms in which characteristics that may represent an advantage are passed from generation to generation by the reproductive success of individuals who carry these characteristics. This reproductive success is linked to sexual selection and the intraspecific competition for mates (Driscoll et al., 2009).

During the process of domestication, natural selection most likely favored animals that had characteristics more amenable to coexisting near humans; for example, animals that were more docile and helped humans in hunting and protected them received resources supplied by humans such as protection and food (Udell et al., 2010).

The artificial selection that was utilized by humans caused changes in several original aspects of dogs (previously wolves), especially in their morphology. Body and head sizes were reduced by the domestication process in dogs as well as in many other species of mammals (Clutton-Brock, 1995). Although dogs have become closer to man and have been domesticated for thousands of years, the phenotypic variation of the dog we know today is more recent, from approximately 3.000 to 4.000 years ago (Driscoll et al., 2009).

The morphology of the current dog, a pet, has diverged in size as well as shape (Savolainen, 2007). Thus many types of dogs have developed with variations in hair color and length, size, eye color, behavior and temperament (Clutton-Brock, 1995). Within this huge variety, currently there are approximately 400 species of dogs, ranging from a little Pinscher to an immense Weimaraner. Some breeds shed less hair and are smaller so that they can be housed in an apartment or small house; some breeds are more docile, some are more or less active, some are better for children or for home and job safety, some are bred to be champions in beauty contests or competitions, etc. More breeds have been developed through the years, either inside laboratories with DNA manipulation or by crossing breeds, always aiming to satisfy humans' vanity and consumerism.

In addition to this manipulation in developing breeds, modifying genetics, morphology and animal characteristics, this process may also lead to several pathological and behavioral problems.

FUNDAMENTAL BEHAVIORS OF DOGS

The influence of human culture on the domestication of dogs evolved into man being the owner and the animal belonging to man, just like any object that can be bought, sold or even traded (Clutton-Brock, 1995). This type of relation, even though replete with affection and positive interactions, began in the domestication stage and remains true today. The domestication process involves changes in morphologic characteristics and also drastically changes some aspects of behavior (Udell et al, 2010). When dogs ("wolf") were wild, they had characteristics and needs that were suppressed and eliminated by human requirements to serve people. According to Hemmer (1990), the animal began to lose its original "perceptual world". Behaviors such as rapid stress reactions for survival in the wild were overshadowed by characteristics such as docility, less fear and more tolerance of stress reactions (Clutton-Brock, 1995).

Domestication has led breeds to underdevelop certain important behavior traits such as intra- and inter-specific social behavior (Udell et al., 2010). The morphologic consequences are strictly related to the consequences of the behavior. Different physical traits exhibited in behavioral signaling among conspecifics were rendered impossible because of morphologic modification. An example is the signals used by wolves through snout, ears, hair, eyes and tail for fundamental social communication. Some breeds today such as pit bulls and Rottweilers have their ears and tails cut off only to satisfy human aesthetics and vanity. Basset hounds, for example, have hearing and signaling problems because of having fallen ears as opposed to breeds that have their ears in a vertical position (Udell et al., 2010). These modifications, in addition to altering or stopping the execution of intrinsic behavior, also cause pain, make animals feel uncomfortable and are often followed by pathologies, hindering their welfare (Broom and Fraser, 2007).

According to Price (1984), domestication affects the animal's behavioral characteristics quantitatively more than qualitatively. The loss of some characteristics in their behavior correlates to the increase of the response threshold for some stimuli. However, a low response threshold may be connected to constant stimulation. An example of this phenomenon of behavioral absence is the lack of an alert reaction to humans. Wild wolves used to feel suspicious or afraid in the presence of humans as would any other wild species; wolves used to have a response threshold of defense or running away from a human when one would approach. As the domestication process developed, instead of being afraid and running, wolves have developed an affective relation in which the defensive and reluctant behaviors have been left behind. Therefore, their alert reactions to humans began to have a different, higher threshold. Another possible explanation for this phenomenon is that it may be a consequence of a combination of handling stimulations, verbal and gestural communication and also physical proximity between humans and dogs, turning an aggressive reaction, what once was a threat, into a situation in which an immediate reaction is not necessary.

Even though the domestication phenomenon has affected the morphological and behavioral aspects of dogs, an extensive behavioral range remains unchanged in wolves such as sociable aspects, sniffing, digging, and burying. Despite the great morphological divergence, the continuation of these behaviors in the *Canis familiaris* species maintains a strong connection to its ancestors, the wolves. Chemical, visual, postural, and vocal communications and social interactions with conspecifics are examples of behaviors that were preserved through evolution. Social standards of behavior, for example, may be considered evolutionary traits that have been maintained in the entire Canidae family (Scott, 1967).

That dogs maintained these behaviors through countless generations shows how much the behaviors characterize the species and thus how essential the behaviors' expression is. Having a gregarious ancestor, the dog has a wide and rich manner of social communication. During puppies' development, in a stage called socialization, and during the reproduction and hunting stages, social behaviors are extremely important for canids, and this intra-specific contact necessity among dogs remains today in all breeds. Social characteristics include visual, odoriferous and vocal contact with conspecificity as the central aspect. A study by Wells and Hepper (1998) of four hundred and seven dogs individually housed in a shelter demonstrated that animals that are provided with visual access to other conspecifics remained positioned in the pen in a manner in which they could see other dogs significantly longer than animals that did not have this access. This finding demonstrates the great need for contact with other conspecifics. Hubrecht et al. (1992) conducted a study with animals from shelters and laboratories; their study showed that animals housed alone remained inactive for a longer time than dogs in groups, being inactive 54%-62% of the time. In addition to being inactive for longer, those housed in groups spend more time displaying active and investigative behaviors such as inspecting the pen by smelling it for a wider variety of smells in a social environment.

In addition to being social, dogs are extremely responsive to their environment, always looking forward to interacting with their surroundings by several methods of communication, mainly smelling (Hubrecht, 2002). The canines' excellent sense of smell is because of the thousands of smell receptors located in their noses, enabling them to detect some odors with a refined precision, odors that are imperceptible to humans. Currently, breeds with better olfaction such as the Labrador retriever and the German Shepherd are trained to identify specific odors of, e.g., drugs, explosives, people after disasters and, even more recently, cancer in some patients (Quignon et al., 2012).

Good vision is essential for dogs to execute some tasks for human benefit, such helping in hunts, guiding blind people, helping the police and communicating with conspecifics and humans. When compared to dogs' vision, humans' must be considered inferior in certain aspects such as acuity, identifying colors, and binocular overlapping. Dogs' vision clearly overshadows humans' vision in areas such as movement detection, field view, ability to distinguish shades of gray and being able to see with little light (Miller and Murphy, 1995).

Like olfaction and vision, vocalization is also of great importance in the communication universe of dogs, either among one another or with humans. Humans' communicating verbally with dogs directly affects their behavior (McConnel, 1990). Understanding the dogs' emotional state may be possible by focusing on the animals' vocalization and posture. The canine vocal repertory is based on 7 types of sounds: (1) *bark* (in alert situations, land defense, individual identification, play and to ease socialization), (2) *howl* (in territorial situations, localizing of group members, individual recognition and coordinated activities such as hunting), (3) *growl* (in defensive situations, danger, play or threat), (4) *yelp* (when suffering from pain or stress), (5) *snore* (nose sound, related to barking), (6) *groan* (under extreme stress) and finally (7) *grunt* (showing pleasure) (Yeon, 2007). These communicative signals are as essential to a mother and puppies as when there is danger or when showing enthusiasm greeting their owners after being left for an entire day.

The positive 12.000-year relation between dogs (wolves) and humans was enabled by the benefits provided by dogs in hunting, chasing and capturing prey and guarding production animals that used to live inside caves and were dangerous (Broom and Fraser, 2007). Since that time, humans and dogs have developed some communication shorthand to create more ease between them

(Savolainen, 2007). Dogs' communications is rich in gestures and expressions as well as in vocalizations and body postures that show such emotions as appreciation, fear, aggressiveness, subordination or dominance, and threat. These signals create ease and comprehension between humans and dogs. Dogs have referential, functional and intentional communication with their owners in several activities such as when they require food (Miklósi et al., 2000). Udell et al. (2010) proposed that communication between dogs and humans depends on two stages; the "Two Stages Hypothesis" is related to the animal's sensitivity to human actions. These interactions are based on the social interactions between dogs and humans as well as the manner in which animals copy humans' body language. Dogs take tips from humans, learning some words from our vocabulary, matching points and recognizing objects. Social stimulations that developed during the natural history of classic conditioning in indoor dogs are crucial to maximizing dogs' quality of interaction (Udell and Wynne, 2008).

As evolution and canine domestication continued, the proximity and cohabitation of human and dog have enabled the sociability between them. Proximity and cohabitation have facilitated clearer communication and have made contact with humans as important to the species as contact with a conspecific.

WELFARE OF DOGS

When dogs were still wild animals living an unhindered life in nomadic human company, they nevertheless had enough freedom to express their natural behavior such as hunting, living among other animals from the identical species, searching for sexual partners, mating, and running; in other words, they were free to express every type of their species' natural behavior. As humans evolved and nomads moved to fixed housing and developed agriculture, domestication became more intense, and dogs' territory became more restricted, close to their owners. When unhindered, animals were exposed to a complex, ever-changing environment in which physical and cognitive requirements were constantly imposed, e.g., avoiding other predators, finding and obtaining food, running through difficult terrain, and defending their territory from other species and groups. Conversely, a restricted environment presents weak natural stimuli and complexity, increased routine, a lack of variety, and processed food. Other factors that result from a captive environment such as homes, animal research, shelters, veterinary clinics, zoonoses control centers, kennels, etc., create a predictable environment, which is important to animal welfare (Swaisgood et al., 2003; Mcphee, 2002).

Broom (1988), one of the pioneers and key researchers of animal welfare, defines welfare as an animal's condition when trying to deal with its environment; this state may vary from very good to very bad. This condition may be accessed by preference tests (Pullen et al., 2010; Vasconcellos et al., 2012) or even using qualitative tests such as the QBA (Qualitative Behavioral Assessment), which integrates behavior expression stiles and signs information, translated into terms linked to emotions (Ruthrford et al., 2012). Several factors may interfere with and impair the welfare of an individual. According to Mcphee (2002), the static conditions of a restrictive environment added to the absence of specific stimuli may result in tedium, an inability to deal with (natural) stress factors, an absence of motivation, and a lack of opportunity to express natural behaviors.

As a consequence of many factors such as noise, dirt, disorder, lack of space, and the need to host several animals, it is normal that environments destined for dogs are restricted and quiet. Environments that do not provide for the social and physiological needs of animals cause serious changes in their behavior and physiology (Hubrecht, 2002) These changes do more than cause stress; in the absence of proper stimuli, behaviors that should be usual are replaced by "unusual" behaviors (Carlstead, 1996).

Abnormal behaviors are those that differ in the manner, frequency, and/or context presented by most of the free-living members of a species (Broom and Johnson, 1993). Examples include coprophagy (feeding of feces), lethargy (inactivity), hyper-aggressiveness, hyper-sexuality, low socialization, self-mutilation, and stereotypies (Boere, 2001; Shepherdson, 1989). Stereotypies is a behavioral disorder that may develop in situations of stress and fear in which the animal feels frustrated at not being able to escape aversive stimuli (Mason, 19991; Carlstead et al., 1993; Shepherdson et al., 2004). Mason (1991) defines stereotypies as standard repetitive behaviors, apparently without any function or goal. These behaviors are physically and temporarily connected to the inappropriate state of the environment. Examples of such behaviors include shaking hands or body, constant chewing, swallowing air and pacing (walking around or from one side to the other frequently and for no reason). Behaviors related to stereotypies in dogs may occur in these forms or, more commonly, may manifest as chasing their own tails in circles or jumping repetitively, which humans see as amusing behavior. All of these behaviors may move animals to high-level pathological states such as depression. Depending on the intensity level of the anomaly, the condition may be irreversible.

Nevertheless, it is essential to emphasize that an individual animal's particularity must not be confused with the general needs of the species. Each dog has a different personality that is derived from its genetic background and life experience (ontogeny) (Hubrecht, 2002). Animals react in different manners to different conditions: scientific experiments, clinical treatments, or even residential environments.

Measures can be taken to avoid and/or minimize the occurrence of abnormal behaviors such as acute or chronic stress. Environmental enrichment can provide comfortable quarters for animals.

An Appropriate Environments for Dogs

Keeping animals in restricted places without changing and compromising their physical and psychological health requires some attention to the basic maintenance of their welfare and comfort. Provision of food, refuge, some control of their living space, physical activity and social contact are some of the basic requirements of captive canines. Dogs are environmentally restricted in many ways: veterinary clinics for treatments, research centers, shelters, kennels or even in their own homes (houses or apartments), which end up being prisons. Inside these confined places are the identical needs that would exist in the wild. With some environmental enrichment techniques, a restrictive environment may become more interactive and provide enough comfort and sufficient welfare.

Environmental Enrichment

The conditions of environmental enrichment of a captive animal consider all the environmental restrictions to which wild, laboratory or production animals or pets are subjected. Providing an interactive and complex environment, socially and physically, and allowing the execution of specific natural behaviors of the species are the main focus of environmental enrichment practices (Carlstead and Shepherdson, 2000; Law et al. 2001; Ellis, 2009).

The practices seek to mimic the natural environment of animals and to provide the behavioral answers they would have when living freely. Practices may range from systematic physical changes in the environment and moving previously existent objects to the introduction of new objects, keeping this new thing in an already known and static place (Genaro, 2005). Utilizing knowledge of the species' biology, simple and creative techniques may stimulate extremely complex behaviors such as hunting. Environmental enrichment can be provided to animals by choice of location, more activities and exploration possibilities, unpredictability, some control of their feeding and social interactions (Damasceno et al., 2012).

Some authors suggest categories for these environmental enrichment techniques. Ellis (2009) and Well (2004), for example, classify them as animated and unanimated. Animated techniques refer to social intra- or inter-specific interactions, and unanimated techniques include changes in the living space's substrate, furnishings, introduction of toys and sensory stimuli. Based on these and other subdivisions, we suggest a classification of five types of environmental enrichment techniques: physical, social, sensorial, feeding and cognitive.

Physical

Physically enriching an environment means modifying and maintaining the enclosure design and infrastructure. Adding or changing substrates, platforms and overall furnishings; access to windows or outdoor places; and the disposition of fixed objects such as bed and food containers are some examples of physical enrichment. All of these techniques purpose to create a moderated complexity in the environment, concomitant to permitted animal space control.

Whatever the animal's captive condition (home, clinic, research center, kennel, etc.) a dog's environment requires some basics such as space, furnishings and a place to exercise.

According to Home Office (1989), a research institution, housing for dogs must follow space criteria based on weight and number of animals. Dogs weighing from 10 to 25 kg, for example, should be housed in a 4.5 m^2 area if alone or 1.9 m^2 per dog when in social conditions, both spaces at least 1.5 m high. Hubrecht (2002) suggests that the size of the pen, beyond being appropriate for animal behavior, also determines how many animals can live together and how many enrichment devices can be introduced.

Introducing new objects or toys, in addition to being an easy technique, can stimulate active and exploratory reactions. Younger animals show a strong response to toys, even after weeks of contact (Hubrecht, 1993). Frequently changing the offered objects and toys is also important for maintaining novelty in the environment (Loveridge, 1998). Suspending objects is an efficient alternative that makes cleaning easier and prevents one member of the group from monopolizing the item (Hubrecht, 1993). In addition to these advantages, when suspended items move, there is more activity as animals interact with them.

The positioning of permanent objects in the environment such as the bed is also relevant. In shelters in which the purpose is the adoption of animals, keeping the bed close to the door may make the dog easier to see, increasing its chances of adoption (Wells, 2004). In veterinary clinics, however, in which the animal is receiving treatment (a stressful situation), the bed must be placed in the back of the room, and it must be well protected so that it can be used by the animal as a refuge.

The use of visual barriers and refuges is extremely important for any captive animal so that it can run away from a possible threat or aversive stimuli, and for that, curtains, boxes, platforms and tunnels are some alternatives (Rochlitz, 2000; Geret et al., 2011; Hubrecht, 2002).

Social

Because dogs are extremely sociable, isolation is highly stressful and can result in behavioral anomalies as well as more frequent and intense vocalizations (Hetts et al., 1992). Animals in research centers used in experiments are often housed alone for infectious pathology control or because they are involved in parasitological studies. In these cases, isolation occurs only for the safety of the animals and the research and only when strictly necessary and for the shortest possible time (Hubrecht, 2002). Dogs require intra-specific interactions; therefore, group housing (Figure 1) is essential. However, despite being extremely important, the choice to house animals in groups must be handled carefully because gathering dogs indiscriminately can cause aggressiveness, and humans who try to interfere in conflict between animals may be hurt (Wells, 2004). There is not a perfect number of dogs per pen as long as there are at least two dogs and sufficient space (Hubrecht, 2002).

Because the domestic dog has interacted with man for thousands of years, the dog-human interaction is part of this species' behavior repertoire. This contact has had a great influence on the maintenance of previously established relations and has created favorable conditions that cause no harm to animals (Tuber et al. 1999); instead, these conditions help in the form of shelters, kennels and research institutions. Beyond creating facilities to improve animal welfare, the existent interspecific interaction with humans has influenced the behavior of the dog and its conspecifics. According to a study by Hubrecht (1993), a small daily increase in dog-human contact may cause great differences in co-specific relations, making them more positive.

Dogs' co-specific and human interactions must be provided in the identical proportion, and none should be excluded or have priority; both are necessary for the dog's welfare (Rooney et al., 2000).

Interaction with other domestic species such as cats (Figure 2) may also be beneficial if the dog is used to it and has had contact with other species during the sensible socialization period (first weeks) (Fox and Stelzner, 1967).

Figure 1.Dogs of undefined breed in physical contact, illustration demonstrating the importance of conspecific interaction for dogs. Picture: Juliana Damasceno.

Figure 2. Inter-specific affiliative interaction between a dog and a cat, both of undefined breed, which have had previous experiences with other species relations. Picture: Juliana Damasceno.

Sensorial

The environmental enrichment sensorial devices are created to stimulate dogs' senses such as olfaction, audition and vision in an attempt to amplify the poor sensorial range of captive animals (Wells, 2009).

Olfaction for dogs is extremely important and constantly used in their perceptual universe. A captive environment generally presents few odoriferous variations, becoming monotonous in this aspect for animals. The introduction of new scents such as of unknown conspecifics, prey, other animals and essences are some alternatives for making the environment odoriferously richer. Toys with scents (Figure 3) also may be introduced or directly placed in a certain section of the environment.

In a study published by Graham et al. (2005a), essences with relaxing properties were introduced such as lavender and chamomile and stimulating essences such as pepper and cloves in a shelter housing 55 dogs. The results of this study showed that essences with calming properties stimulate docile behaviors and activities that suggest relaxing, such as resting. This behavior is considered beneficial to animals' welfare and facilitates animal adoption. Conversely, the pepper and clove odors stimulated alert behaviors and vocalization, abetting the cognitive and cerebral stimulation of these animals. Odoriferous enrichment is a practical, quick and effective technique.

Although dogs do not have an accurate visual system, television images and mirrors can be good visual enrichment. In a study involving 50 dogs in a shelter, Graham et al. (2005b) had significant results when they presented images of conspecifics and humans in movement on televisions placed in front of pens.

Figure 3. Labrador retriever interacting with toy containing a lavender scent, an example of odoriferous enrichment in the environment. Picture: Juliana Damasceno.

Classical music may be a favorable option for auditory stimulation of most beleaguered animals (Wells, 2004). Nature sounds and same-species sounds can also be an interesting alternative.

Feed

A dog's diet must be balanced and nutritious according to the animal's age and weight. A dietary environmental enrichment technique involves variations in the presentation, items and shape of dog food.

In animals' extensive behavioral repertoire, food deserves to be highlighted because in the wild (freely living), mammals spend the majority of their time looking for, capturing and consuming their food (Jenny and Schmid, 2002). Several behaviors are involved in this activity that are absent in a captive environment because of regular and easy access to food.

Some manners of enriching without abruptly interfering in animals' daily diet include different manners of presenting food such as whipped, frozen (in hot weather), spread (in clean places), suspended or hidden inside toys or in the environment itself (Law et al., 2001; Ellis, 2009). In this manner, animals are stimulated to look for their meals and spend more time and energy during meals, which does not occur when food is always presented in the identical manner, time and disposition.

Items found in markets that are different from the usual ones such as snacks, pasty food and bones (Figure 4) can also be offered in creative manners, always taking care to avoid nutritional problems and excess weight.

Figure 4. Dog of undefined breed interacting with bones that are easily found in pet shops, an example of a practical and effective feed enrichment device. Picture: Juliana Damasceno.

Cognitive

Cognitive, or occupational, enrichment stimulates the animal's ability to solve problems. Puzzles with food rewards and hidden food are effective cognitive techniques. Puzzles containing food are extrinsic reinforcement, stimulating animals to solve problems so that they can have a reward. The animal's response will be to try to solve the problem; receiving a reward increases the probability that a certain behavior will be repeated, thus decreasing the tendency of habituation to the enrichment item (Tarou and Bashaw, 2006). Actually some devices such as treat balls (Figure 5) and puzzle-feeders (Figure 6) are available in pet stores and are great options for cognitive enrichment.

Although enrichment techniques are classified into different categories, actual stimulus division rarely occurs. Generally a technique includes one or more types of enrichment. For example, the simple introduction of a new object containing a food reward inside is simultaneously applying the physical, the sensorial (olfactory and visual) and also feed enrichment types. However, these divisions are efficient for explanation and help in understanding animal behavior and how to enrich their environment.

Figure 5. A Jack Russell terrier interacting with a PetGames® treat ball. Picture: Dalton Alexandre Ishikawa.

Figure 6. An Australian cattle dog interacting with a PetGames® puzzle-feeder. Picture: Dalton Alexandre Ishikawa.

The importance of these techniques for animals' behavioral, physical and psychological benefits is shown in the literature in numerous studies. These studies are related to techniques for reducing unusual behaviors (Resende et al. 2011; Quirke and Riordan, 2011; Sherwin et al., 1999), reducing inactivity and increasing natural behaviors (Ellis and Wells, 2008; Wells and Egli, 2004; Schipper et al., 2008; Morimura and Ueno, 1999) as well as improvements in reproductive rates (Moreira et al., 2007; Carlstead and Shepherdson, 1994).

Even though these techniques are easy to apply and efficient activities, we must take care when applying enrichment techniques because if they are not properly executed or if they are executed without sufficient knowledge of the species' biology, there are risks such as injuries and stress that nullify the goals of the techniques. Mellen and MacPhee (2001) suggest that for efficient enrichment and to achieve expectations, those responsible for the application must follow five steps: (1) setting goals, (2) planning, (3) implementing, (4) documenting, and (5) re-adjusting. If these steps are followed, the probability of efficiency and safety is high, and it will be clear whether the animals are benefitting from the applied technique.

Enriching a dog's environment will render the animal more active and psychologically healthy, present behaviors that are typical of its species and lower the probability of aggressive or defensive behaviors. The maintenance of activities and the addition of items to stimulate natural behaviors and avoid stress will help to ensure that an animal's physical and psychological health has an excellent chance of remaining at satisfactory levels.

REFERENCES

American Veterinary Medicine Association. (2012). *American Veterinary Medicine Association.2012 U.S. Pet Ownership & Demographics Sourcebook.* Schaumburg, II: American Veterinary Medicine Association. https://www.avma.org/KB/Resources/Statistics/Pages/Market-research-statistics-US-Pet-Ownership-Demographics-Sourcebook.aspx

Broom, D. M. (1988).The scientific assessment of animal welfare. *Applied Animal Behaviour Science, N°20*, 1988, 5-19, 0168-1591

Broom, D. M. & Johnson, KG. (1993). *Stress and animal welfare.* (1st Ed). Dordrecht, Netherlands: Kluwer Academic Publishers.

Broom, D. M. & Fraser AF. (2007). *Domestic animal behavior and welfare.* (4th Ed). Oxfordshire, UK: CABI.

Boere, V. (2001). Environmental enrichment for Neotropical primates in captivity: a review. *Ciência Rural, N°31*, 2001, pp. 451-460, 0103-8478.

Carlstead, K., Brown, J. & Seidensticker, J. (1993). Behavioral and adrenocortical responses to environmental changes in leopard cats (*Felisbengalensis*). *Zoo Biology. N 12*, 1993, 321-331, 1098-2361.

Carlstead, K. & Shepherdson, D. J. (1994). Effects of environmental enrichment on reproduction. *Zoo Biology. N 13*, 1994, 447–458, 1098-2361.

Carlstead, K. (1996). Effects of captivity on the behavior of wild mammals. p 317-333. In: Kleiman, D., Allen M., Thompson NM, Lumpkin, S. (Eds) *Wild Mammals in Captivity.* (1st Ed, pp. 317-333). Chicago, Illinois: University of Chicago Press.

Carlstead, K. & Shepherdson, D. J. (2000). Alleviating stress in zoo animals with environmental enrichment. In: Moberg GP., Mench JA. (Eds). *The biology of animal stress: basic principles and implications for animal welfare.* (1st Ed, pp. 337-354) New York, USA: CABI Publishing.

Clutton-Brock, J. (1992). The process of domestication. *Mammal Review. N° 22*, 1992, 79-85, 1365-2907.

Clutton-Brock, J. (1995). Origins of the dog: domestication and early history. In: Serpell J. (Eds) *The domestic dog: its evolution, behaviour and interactions with people.* (1st Ed, pp. 7-19) Cambridge, UK: Cambridge University Press.

Damasceno, J., Adania C. H. & Genaro, G. (2012). Bem-estarparaani maisemcativeiro. BVS- VET Biblioteca Virtual emMedicinaVeterinária e Zootecnia. http://bvsvet.blogspot.com.br/2012/10/bem-estar-para-animais-em-cativeiro.html

Davis, S. J. M. & Valla, F. R. (1978). Evidence for domestication of the dog 12,000 years ago in the Natufian of Israel. *Nature*, 1978, *276*, 608-610.

Driscoll, C. A., Macdonald, D. W. & O'Brien, J. (2009). From wild animals to domestic pets, an evolutionary view of domestication. *Proceedings of the National Academy of Sciences of the United States of America (PNAS). N°106*, 2009, 9971-9978, 0027-8424.

Ellis, S. L. H. & Wells, D. L. (2008). The influence of visual stimulation on the behaviour of cats housed in a rescue shelter. *Applied Animal Behaviour Science, N°113*, 2008, 166–174, 0168-1591.

Ellis, S. L. H. (2009). Environmental Enrichment: Practical strategies for improving feline welfare. *Journal of Feline Medicine and Surgery. N°11*, 2009, 901-912, 1532-2750.

Fox, M. W. & Stelzner D. (1967). The effects of early experience on the development of inter and intraspecies social relationships in the dog. *Animal Behaviour. N°15*, 1967, 377-386, 0003-3472.

Genaro, G. 2005. Gatodoméstico – comportamento & clínica veterinária. *Revista MEDVEP*. N° 3, 2005, 16-22, 1678-1430.

Geret, C. P., Cattori, V., Meli, M. L., Hofmann-Lehmann L. & Lutz, H. (2011). Housing and care of laboratory cats: from requirements to practice. *SchweizerArchvfürTierheilkunde. N°153*, 2011, 157-164, 1664-2848.

Graham, L., Wells, D. L. & Hepper, P. G. (2005a). The influence of olfactory stimulation on the behaviour of dogs housed in a rescue shelter. *Applied Animal Behaviour Science. N° 91*, 2005a, 143-153, 0168-1591.

Graham, L., Wells, D. L. & Hepper, P. G. (2005b).The influence of visual stimulation on the behaviour of dogs housed in a rescue shelter. *Animal Welfare. N°14*, 2005b, 143-148, 0962-7286.

Home Office. (1989). *Animals (Scientific Procedures) Act 1986. Code of Practice for the Housing and Care of Animals Used in Scientific Procedures.* Her Majesty's Stationery Office, London, UK Full Text: http://www.homeoffice.gov.uk/animact/hcasp.htm.

Hemmer, H. (1990). *Domestication: the decline of environmental appreciation.* Cambridge, UK: Cambridge University Press.

Hetts, S., Clark, J. D., Calpin, J. P., Arnold, C. E. & Mateo, J. M. (1992). Influence of housing conditions on beagle behaviour. *Applied Animal Behaviour Science. N° 34*, 1992. 137-155, 0168-1591.

Hubrecht, R. C., Serpell, J. A. & Poole, T. B. 1992. Correlates of pen size and housing conditions on the behaviour of kennelled dogs. *Applied Animal Behaviour Science. N°34*, 1992, 365–383, 0168-1591.

Hubrecht, R. C. (1993). A comparison of social and environmental enrichment methods for laboratory housed dogs. Applied *Animal Behaviour Science. N°37*, 1993, 345-361, 0168-1591.

Hubrecht R. (2002). Comfortable quarters for dogs in research institutions. In: Reinhardt V., Reinhardt A. (Eds) *Comfortable Quarters for Laboratory Animals*, (9th Ed, pp.56-64). Washington, DC: Animal Welfare Institute. www.awionline.org/pubs/cq02/Cq-dogs.html.

Jenny, S. & Schmid, H. (2002). Effect of feeding boxes on the behavior of stereotyping amur tigers (*Pantheratigrisaltaica*) in the Zurich Zoo, Zurich, Switzerland. *Zoo Biology. N°21*, 2002,573- 584, 1098-2361.

Law, G., Graham, D. & Mcgowan, P. (2001). Environmental enrichment for zoo and domestic cats. *Animal Technology. N°52*, 2001, 155-163, 0264-4754.

Loveridge, G. G. (1998). Environmentally enriched dog housing. Applied *Animal Behaviour Science. N°59*, 1998, 101-113, 0168-1591.

Mason, G. J. (1991). Stereotypies: a critical review. *Animal Behaviour. N°41*, 1991,1015–37, 0003-3472.

McConnell, P. B. (1990). Acoustic structure and receiver response in domestic dogs, *Canisfamiliaris. Animal Behaviour. N° 39*, 1990, 897-904, 0003-3472.

Mcphee, M. E. (2002). Intact Carcasses as enrichment for large felids: effects on on–and off exhibit behaviors. *Zoo Biology. N° 21*, 2002, 37-47, 1098-2361.

Mellen, J. & MacPhee, M. S. (2001). Philosophy of Environmental Enrichment: Past, Present, and Future. *Zoo Biology. N°20*, 2011, 211-226.

Miller, P. E. & Murphy, C. (1995). Vision in dogs. *Journal of the American Veterinary Medical Association (JAVMA). N° 207*, 1995, 1623-1634, 0003-1488.

Miklósi, A., Polgárdi, R., Topál, J. & Csányi V. (2000). Intentional behaviour in dog-human communication: an experimental analysis of "showing" behaviour in the dog. *Animal Cognition.* N°3, 2000, 159-166, 1435-9456.

Moreira, N., Brown, J. L., Moraes, W., Swanson, W.F. & Monteiro-Filho, E. L. A. (2007). Effect of housing and environmental enrichment on adrenocortical activity, behavior and reproductive cyclicityin the female tigrina (*Leopardustigrinus*) and margay (*Leoparduswiedii*). *Zoo Biology. N° 26*, 2007, 441-460, 1098-2361.

Morimura, N. & Ueno, Y. (1999). Influences on the feeding behaviour of three mammals in Maruyama Zoo: bears, elephants, and chimpanzees. *Journal of Applied Animal Welfare Science. N°2*, 1999. 169-186, 1532-7604.

PFMA. (2012). PFMA Pet Food Manufacturers' Association, 2012. http://www.pfma.org.uk/pet-population/.

Price, O. E. (1984). Behavioral aspects of animal domestication. *The Quarterly Review of Biology. N° 59*, 1984, 1-32, 0033-5770.

Pullen, A. J., Merill, R. J. N. & Bradshaw J. W. S. 2010. Preferences for types and presentations in kennel housed dogs. *Applied Animal Behaviour Science. N°125*, 151-156, 0168-1591.

Quignon, P., Robin S. & Galibert, F. 2012. Canine olfactory genetics. In: Ostrander E.A., & Ruvinsk A. (Eds). *The genetics of the dog* (2nd ed., 375- 393). Wallingford, UK: CABI Publishing.

Quirke, T. & O'Riordan, R. M. (2011). The effect of a randomized enrichment treatment schedule on the behavior of cheetahs (*Acinonyxjubatus*). *Applied Animal Behaviour Science. N° 135*, 2011, 103-109, 0168-1591.

Resende, L. S., Gomes, K. C. P., Andriolo, A., Genaro, G., Remy, G. L., & Ramos, V. A. (2011). Influence of cinnamon and catnip on the stereotypical pacing of oncilla cats (*Leopardustigrinus)* in Captivity. *Applied Animal Welfare Science. N°14*, 2011, 247-254, 1532-7604.

Roots, C. (2007). *Domestication.* Westport, USA: Greenwood Press.

Rutherford, K. M. D., Donald, R. D., Lawrence, A. B. & Wemelsfelder F. (2012). Qualitative behavioural assessment of emotionaly in pigs. *Applied Animal Behaviour Science. N°139*, 2012, 218-224, 0168-1591.

Savolainen, P. (2007). *Domestication of dogs.* In: Jensen P. (Eds), The behavioural biology of dogs. (1st Ed, 21-37) Wallinford, UK: CABI Publishing.

Scott, JP. (1967). The evolution of social behavior of dogs and wolves. *American Zoologist, N°07*, 2007, 373-381, 0003-1569.

Sherwin, C. M., Lewis, P. D. & Perry, G. C. (1999). The effects of environmental enrichment and intermittent lighting on the behaviour and welfare of male domestic turkeys. *Applied Animal Behaviour Science. N° 62*, 1999, 319- 333, 0168-1591.

Rochlitz, I. (2000). Recommendations for the housing and care of domestic cats in laboratories. *Laboratory Animals. N°34*, 2000, 1-9, 1758-1117.

Rooney, N. J., Bradshaw, J. W. S. & Robinson, I. H. (2000). A comparison of dog–dog and dog–human play behaviour. *Applied Animal Behaviour Science. N° 66*, 2000, 235-248, 0168-1591.

Swaisgood, R. R., Ellis. S., Forthman, D. L. & Shepherdson, D. J. (2003). Commentary: improving well-being for captive giant pandas: theoretical and practical issues. *Zoo Biology, N° 22*, 2003, 347–354, 1098-2361.

Shepherdson, D. J. (1989). Improving animals' lives in cap- tivity through environmental enrichment.   91- 100 in Euroniche Conference Proceedings. Close, B.S., Dolins, F., Mason, G., eds. London, Humane Education Centre.

Shepherdson, D. J., Carlstead, K. C. & Wielennowski N. (2004). Cross-institutional assessment of stress responses in zoo animais using longitudinal monitoring of faecal corticoids an behaviour. *Animal Welfare, N°13*, 2004, 105-113, 0962-7286.

Schipper, L. L., Vinke, C. M., Schilder, M. B. H. & Spruijt, B. M. (2008). The effect of feeding enrichment toys on the behaviour of kennelled dogs (*Canisfamiliaris*). *Applied Animal Behaviour Science. N°114*, 2008, 182- 195.

Tarou, L. R. & Bashaw, M. J. (2007).Maximizing the effectiveness of environmental enrichment: suggestions from the experimental analysis of behavior. *Applied Animal Behaviour Science. N°102*, 2007, 189-204.

Tuber, D. S., Miller, D. D., Caris, K.A., Halter, R., Linden, F. & Hennessy, M. B. (1999). Dogs in animal shelters: problems suggestions and needed expertise. *Psychological Science. N°10*, 1999. 379–386, 0956-7976.

Udell, M. A. R. & Wynne, C. D. L. (2008). A review of domestic dogs' (*Canisfamiliaris*) human-like behaviors: or why behavior analysts should stop worrying and love their dogs. *Journal of the experimental analysis of behavior. N° 89*, 2008, 247-261, 1938-3711.

Udell. M. A. R., Dorey, N. R. & Wynne C. D. L. (2010). What did domestication do to dogs? A new account of dog's sensitivity to human actions. *Biological Reviews. N°85*, 2010, 327-345, 1469-185X.

Valla, F. R. (1990). Le Natufien: uneautrefacon de comprendre le monde? *Mitekufat Haeven, Journal of the Israel Prehistoric Society. N° 23*, 1990, 171-175.

Vasconcellos, A. S., Adania, C. H. & Àdes, C. (2012).Contrafreeloading in maned wolves: Implications for their management and welfare. *Applied Animal Behaviour Science. N°140*, 2012, 85-91, 0168-1591.

Vilà, C., Savolainen, P., Maldonado, J. E., Amorim, I. R., Rice, J. E., Honeycutt, R. L., Crandall, K. A., Lundeberg, J. & Wayne, R. K. (1997). Multiple and ancient origins of the domestic dog. *Science. N°276*, 1997, 1687-1689, 1095-9203.

Wayne, R. K. & Vilà, C. (2001). Phylogeny and origin of the dog. In: Ruvinsky A., Sampson J. (Eds), *The genetics of the dog*. (1st Ed, pp. 1 - 14) New York, USA: CABI Publishing .

Wells, D. L. & Hepper, P. G. (1998). A note on the influence of visual conspecific contact on the behaviour of sheltered dogs. *Applied Animal Behaviour Science. N° 60*, 1998, 83-88, 0168-1591.

Wells, D. L. (2004). A review of environmental enrichment for kennelled dogs, *Canisfamiliaris. Applied Animal Behaviour Science. N° 85*, 2004. pp. 307-317, 0168-1591.

Wells, D. L. & Egli, J. M. (2004). The influence of olfactory enrichment on the behaviour of black-footed cats, *Felisnigripes. Applied Animal Behaviour Science. N°85*, 2004. 107-119, 0168-1591.

Wells, D. L. (2009). Sensory stimulation as environmental enrichment for captive animals: a review. *Applied Animal Behaviour Science. N° 118*, 2009, 1-11, 0168-1591.

Yeon, S. C. (2007).The vocal communication of canines. *Journal of Veterinary Behavior Clinical applications and Research. N°2*, 2007, 141-144, 1558-7878.

In: Dogs
Editors: K. M. Cohen and L. R. Diaz

ISBN: 978-1-62808-530-3
© 2013 Nova Science Publishers, Inc.

Chapter 3

MARKETING AND THE DOMESTICATION OF DOGS

José Antonio Soares[1,], André Luiz Baptista Galvão[2],*
Amanda Leal Vasconcellos[2], Elzylene Léga Palazzo[3],
Manuela Cristina Vieira[2], Thais Rabelo dos Santos[2],
Rodrigo Rabelo dos Santos[2]
and Katia Denise Saraiva Bresciani[2,4,†]

[1]FKB, Fundação Karnig Bazarian, Faculdades Integradas de Itapetininga,
Itapetininga, São Paulo, Brasil.
[2]UNESP, Universidade Estadual Paulista, Faculdade de Ciências Agrárias
e Veterinárias de Jaboticabal, Jaboticabal, São Paulo, Brasil.
[3]FAFRAM, Faculdade"Dr. Francisco Maeda", Ituverava,
São Paulo, Brasil.
[4]UNESP, Universidade Estadual Paulista, Faculdade de Medicina
Veterinária de Araçatuba, São Paulo, Brasil.

---

[*] E-mail: prof.soares@uol.com.br
[†] E-mail: bresciani@fmva.unesp.br

ABSTRACT

If we think about Marketing with ethics and professionalism, we can see it as an effective tool in the continuous pursuit of quality related to the provision of veterinary services. What we see today is the usual loss of professionals of this area, whose concern is limited to caring for small animals, especially dogs. We cannot lose sight of the business, since an enterprise is founded to make profit; obviously the small animal care has a huge importance, but the company needs to survive and grow so that new services can be offered to satisfy those consumers and customers. The survival of those companies is really possible, with profit and ethics, focusing on the animal's health and welfare.

INTRODUCTION

Dogs were domesticated, in essence, to provide territorial security and support in services and hunts. Currently, its reproduction is often used as a way of obtaining revenue and they are even considered as family members. In marketing terms, small animals have marked global relevance, having business in this sector increased considerably in the recent decades, as can be evidenced by the steady growth of its food, pharmaceuticals, drugs and special diets markets (Bernasconi, 2007).

In particular, within the Pet Shop sector, there is a development and specialization of food and pharmaceutical products according to the dog's needs. Thus, there is a wide variety of services for a consumer avid for news and information (Bernasconi, 2007).

Thereby, in the case of canine nutrition, numerous niches are ascending to privileged positions in the PET global market, in the form of fresh, cold, raw and organic food, raw organic, grain free, with human quality standard ingredients, natural or even exotics, "super premium", "ultra premium", homemade meals enriched with supplements, nutrition based on meat (meat-centric), proteins (protein focused), and niche diets, for example, for senile animals, athletes, puppies in training, as well for oral, skin, hair, intestinal tract, urinary health or for each condition of race and illness condition (Saad and France, 2010).

According to the National Manufacturers Association of Products for Pets (Anfalpet), Brazilian market for pets, in 2010, was worth 11 billion reais. 66% correspond to the sale of food for pets and 20% to the sector services. This is a really promising market. Worldwide, the industry earned US$ 76 billion in 2010.

Marcos Gouvea de Souza, GS and MD - Gouvea de Souza CEO, considers good the perspectives for the pet shops segment in Brazil. In 2010, this market grew 8.5% in comparison to the previous year (2009) and the segment had revenues, at the retail, that exceeded US$ 11.3 billion, with a real growth of 4.5% in 2011.Thus, it seems that expansion will continue growing in a percentage higher than the GDP growth for the coming years, with a estimated number of 25,000 pet shops in the country.

The innovation, with differentiated products, is a pet shops market trend, such as nail polish and soft drinks, beauty salons for pets with new types of baths, trims and drying, with heavy use of imported products and delivery services. On the luxury issue, Brazilian pet market still has much to grow. A survey conducted by the website "WebLuxo" revealed that only 5.5% of the Brazilian market is composed of more expensive products, and, among them, the most sold are collars, clothes, drinking fountains and houses.

The United States leads the world pet market, followed by Brazil. According to data from Anfalpet, this last country has structure and production capacity to be also the second largest exporter of goods on the segment, with US$ 4 billion per year.

Brazil has 98 million pets. According to Antonio Braz, IBGE (Brazilian Institute of Geography and Statistics) analyst, the share of spenses on pets achieve the 0.7% of the budget. So it is good to pay attention to the products and services trends and to the basic maintenance rules of a pet shop or veterinary clinic.

For implementation and supply of products and services, marketing strategies are important, and among them: price, product, distribution (sales point) and promotion, allow to place these products and services with emphasis on establishments and, by this way, to promote sales, which Kotler (2000) titled as Marketing Mix or 4 P's, as called McCarthy (1996).

Ogden (2002) features several forms of communication that can be used to highlight and bring to light such products with the use of: banners, signs, blimps, stickers, posters, mobiles, brochures, stopper, tasting counters, balloons, shop windows, billboards, displays and mechanical devices. Smells, sounds and special decoration can also compose pet retail environment. Social networks are widely used in customer's education and awareness about the consumption of sold products, as therapeutic diets, risks and care of obese animals, oral health and guidance on the vaccination schedule. In this context, the practice of the IMC (Integrated Marketing Communications) concept can be applied, in order to define the Marketing and Communications objectives to be achieved for the products and brands in object and the investments to be applied (Pinheiro and Gullo 2009).

Currently, the rising number of pets in the society and the need to maintain animal welfare and public health has triggered scientific researches for the development of veterinary drugs.

Decree number 5053 (April 22, 2004), regulates the Decree-Law number 467 (February 13, 1969), as provided in its article 12, and approves the Regulation of Inspection of Products for Veterinary Use and of Establishments that Produce or Sell it. It is jurisdiction of The Ministry of Agriculture, Livestock and Supply (MAPA) to enact additional regulations relating to manufacturing, quality control, selling and use of veterinary products, and other relevant actions for the normalization of the Regulation, including those approved under the Southern Common Market (Mercosur) (MAPA, 2012).

Therefore, after the aforementioned Decree, occurs the obligatoriness of clinical trials, with laboratory and field testing for every active ingredients and raw materials, according to the indicated type and dose and before the start of products sales. So, we can observe a standardization in the development of veterinary pharmaceuticals, through a more rigid quality control.

The conscious use of drugs, vaccines and supplements, are part of preventive and healing veterinary medicine, helping to maintain animal health and, consequently, human health. It is needed to mention that the misuse of drugs can trigger resistance by pathogens besides undesirable side effects.

Being a competent veterinarian in properly applying Marketing techniques is, today, prerequisite for the development of a strong business. A competent professional does not violate the Veterinary Medicine ethics when using marketing techniques to meet the needs of their patients. An adequate approach will cause customers to invest in their furry companion's health (Pereira, 2010a; Pereira, 2012c).

Marketing is not a set of traps to mislead the consumer or to induce the customer to spend on what he doesn't want or on something that he does not need. A major goal of marketing techniques is to have customers for life, ie make them loyal, so it's important to get their confidence by providing them good experiences and, especially, strengthening the commitment to their pet (Pereira, 2010a.). Therefore, the Marketing constitutes a system of activities that interact, designed to plan, fix the sale price, propagate and deliver services that comply those customers and patients, always recognizing and meeting their requirements, needs and wishes (Pereira, 2012c).

The practice of veterinary medicine consists to prevent and treat animal diseases, among other duties. However, in the Marketing view it consists to offer solutions to the pet's problems. Health and animal welfare will always be the objectives of the used marketing techniques, which implies in making the dog owner to understand what is being done, existing great importance in the communication with the client, to the appreciation of their commitment (Pereira, 2012c).

Today's competitive market requires the veterinarian to be economically active, which means not having to sell what the animal doesn't need, but offer everything the Veterinary Medicine and industries of related products have to meet the needs of each animal (Pereira, 2009a).

The main idea is that the veterinarian offer everything the dog needs and provide specific health programs, veterinary medicines, proper feed, proceeding in a way that the patient receives the very best, satisfying owners with the quality of its treatment (Pereira, 2009a).

1. DISCLOSURE TO THE PUBLIC ABOUT THE SERVICES PROVIDED AND AVAILABLE IN YOUR BUSINESS

There are several reasons that can bring new customers to a Veterinary Clinic: (a) the purchase of the first pet, (b) the provision of differentiated and specialized services in different areas of veterinary medicine, (c) owners unhappy with the service provided in another establishments, (d) easy location and access to the business, and (e) recommendations from friends and colleagues. However, potential customers inform themselves about the company among friends, colleagues, clinic sites, read online reviews and check pages in social networks (Felsted and Brakke, 2013). The "word of mouth" marketing works very well in these cases.

In the search for a veterinary service, the customer wants quality, safety and accessibility adequate to his animal needs and when choosing a veterinary clinic or hospital, the customer believes that he will have the expected service. So the veterinarian has to exceed the customer's expectations and offer a service better than expected, differentiating from other competitors and interacting with the customers (Pereira, 2009b; Pereira, 2010a). Delighting the customer can lead to loyalty.

Many owners have strong emotional involvement with their dogs. These customers should be heard and treated with patience and understanding, by the

Veterinary Clinic crew, in a standardized, affectionate and respectful way (Pereira, 2009b; Pereira, 2010a; Pereira, 2012b; Gioso et al., 2011c). Professionals who listen to their client will have a great opportunity to differentiate themself from the others, so is recommended to the professional to saturate his company with the customer voice (Whiteley, 1992). Through these relationship practices, ties with the customers can be established and they may be likely to return and fulfill the asked recommendations (Felsted and Brakke, 2013).

An important point is to make business cards available to new customers at the end of the consultation, with all the veterinary clinic data, such as address, phone, email and social networking pages (Felsted and Brakke, 2013).

Professionals must use available internet and social networks resources as ways to attract customers by creating a website, blog or social network page of the business, with the clinic brand, adding photos of its space and crew, presenting good appearance, which helps to publicize the service. Initially the professional usually publicize the website for his loyal and regular customers, asking the same for authorization to disseminate photos and stories of their dogs, so his customers will be able to make comments and testimonials on them. Through social networking, you can publicize services and resources to diagnose and educate customers about nutrition and obesity, or even about the importance of different diets applied for different pathologies. It would be interesting to make available a link to customers or visitors with information about the need of preventive care of their dogs (Felsted and Brakke, 2013).

There may also be, on the website, an online pharmacy, with medication information, presenting human products that are toxic to dogs and promoting the dissemination of veterinary pharmaceutical products available in the clinic. Another smart option is the inclusion of educational videos on various topics, such as animal welfare or about the importance of dental cleaning for the establishment clients (Materni and Tumblin, 2013). By creating this page, it is possible to keep partnerships with the pharmaceutical companies, such as laboratories, pharmacies and veterinarian stores, seeking sponsorship and real promotion of their brands and products, leading to lower investment in this type of advertising of the company, besides expanding the network, increasing the reliability of the customers regarding the company.

A web page is an excellent opportunity of business promotion, since the internet is the greatest information source nowadays. There are, already, very original pages, which contain information such as offered products and services, operation hours, address, phone, curiosities about the animals races and even scheduling of bathing and grooming (Salvador, 2007). However, the page author must be careful and clear with its content and utilize easy access keywords for its articles (Schaible, 2012).

Marketing strategies are numerous and aim to retain good customers and acquire new ones. Thereby mixtures of various media are valid; direct mail, e-mails, phone calls and explanatory brochures, being integrated communication the most effective. In this aspect, the professional cannot be inconvenient, sending many emails to the customers and calling too much for the animal owner. A call from the vet after a consultation or performed surgical procedure, to inform the condition of the dog is recommended; not to advertise products (Gioso et al., 2011c).

Another way to differentiate the establishment is to show to the public, support to actions with social involvement. Veterinary hospitals in the U.S. have been prepared in gathering supplies for people and animals to local charities institutions, being given a discount on services to the customer when bringing a donation (Tudor, 2013).

Suggested marketing actions constitute a proposal which everyone wins. The client gains in economy and satisfaction of having a healthy dog; the patient will benefit the most, gaining in health and wellness. And the veterinarian, gain in offering a quality service for an extended period, maintaining economic viability throughout the year (Pereira, 2007).

What is important is not the quantity but the quality of communication (Pereira, 2010b). It is extremely important that the client understands the quality of treatment offered, being vital to educate and inform the customer of the veterinarian role in his pet life, using appropriate Marketing techniques (Pereira, 2010c).

2. THE RECEPTION

The veterinarian should wish the owner and the patient to feel comfortable, in a safe environment (Stewart, 2008). This business establishment must provide accommodations, facilities and suitable environment for patients and owners, as well as water, tea, coffee, sweets and biscuits and the availability of a presentable toilet. In situations of rain, it is interesting for clients having a member of your staff, with an umbrella, in attendance to help clients when leaving and entering the clinic towards their vehicles (Stewart, 2008; Barforoush, 2012).

In intelligent scheduling, whenever possible, the appointments are scheduled for patients according to species. Cats who do not share their home with dogs may show anxious and stressed when taken to Veterinary Clinic and therefore should be placed in separate and private environment, preferably in a shipping box in the isolated room (Stewart, 2008).

Preferably, patients should not move freely throughout the establishment, as they, once free, can cause accidents, insult other animals, escape or even spread diseases (Stewart, 2008).

In the personal relation, each staff member greets customers by owners' and dogs' names, thanks them the visit to the Clinic and takes a photo of the dog for medical record. Regardless of who is at the front desk, customers receive the same warm greeting and it becomes a trademark of its practice (Stewart, 2008).

For Las Casas (2006), it costs five times more to win a new customer than to keep an existing customer. Thus, small details make the difference and should not be neglected, such as hygiene around the place, ease of approach, operating hours, appointment scheduling, always treat the customer and patient by name, full explanation of procedures and full billing (Pereira, 2012B). Some clinics have a separate reception room, where the owner can wait while his pet is attended or undergo any surgical procedure (Pereira, 2009b). It is important to remember that the veterinary service can be of high quality but, in a matter of seconds, one careless placement of a crew member can put everything away.

3. THE SERVICE

Over the years, dog owners have become more demanding as they are learning how to spend money on what really matters. Thereby, in the clinical practice, this client studies and examines their chances of spending in relation to the diagnosis methods and treatment forms, questions, discusses and refuses services if the explanations are not persuasive or they do not feel safety from the veterinarian (Tracey 2011). The client must have an overview of the service that is being provided.

The more education and respect the clinic staff shows to the owner and patient, the better its reputation (Tracey, 2011). The way the customer is treated makes the difference against the competition. It is worth mentioning that the patient needs must be attended and the customer may be psychologically unstable then, a gesture and a well-placed word can make the difference.

It is known that simply speaking may not be sufficient. So, the vocabulary should be simple, non-technical and at the end of the service, the professional has to be sure that the customers understood his recommendations (Pereira, 2008).

It is important the veterinarian to conduct a careful physical examination of the patient and discuss what is happening with the dog owner, explaining clearly and honestly the risks involved around the disease in question, options for diagnosis and treatment modalities, accurately, prioritizing less aggressive treatments. The pet owner, at this moment, reaffirms its confidence in the professional, regarding the investment of his time and money (Tracey, 2011).

After appointment, the veterinarian must give the examinations results to the owner, as well as monitor the progress of the therapy applied to the patient, scheduling returns, calling and/or making contact by e-mail (Tracey, 2011).

Throughout the treatment procedure, the owner has to be aware of what his pet is feeling. It will help him to decide what should be done immediately, what is priority and what can be expected, so it establishes how and what he can and should invest (Becker, 2013).

4. EXPANDED SERVICE

Specialized veterinary medical clinics make much difference. Thus, a clinic may, for example, have a specific ambulatory for immunization, another for ectoparasites control (where the product is sold and applied), operating room, intensive care unit (ICU), clinical laboratory, hospital room and even a room with electrocardiogram (Pereira, 2011). Thus, it is not necessary to have a mega structure to succeed in the market; you need organization, excellence in service and dedication to get recognition from the customers (Lobato, 2010).

Medical specialties and programs are designed to meet the needs of each client and patient in particular. Two examples of programs can make the difference in the competitive Pet world market.

The first of these programs is Pediatrics. Guidelines to customers should be made available in the first consultation (Pereira, 2005), with the settlement of a prophylactic, infectious and parasitic schedule, as well as the elucidation of the balanced nutrition importance for growth and dog development (Pereira, 2005). If the client joins the program, he will start a relationship that will last for years, if well administered, leading to loyalty of the owner to the veterinarian (Pereira, 2005).

The second program is the Geriatrics. When the dog reaches the age of seven, starts to become an aged animal, requiring special care to stay healthy and have a increased life expectancy (Pereira, 2012a).

The main idea is that, to meet all the needs of an elderly dog, it is appropriate to offer a complete geriatric program. The vet, then, conducts a survey of registered animals with seven years or more and get in contact with the customer to explain about senility and demonstrate concern for the health of his patient. From this alert, the owner realizes that the dog is aging and this is crucial for its accession to the geriatric program (Pereira, 2012a).

The geriatric program comprises a standard format, and should not be confused with package, because a dog has special needs depending on sex, race, previous morbid state, among other factors. The program scope should adapt to each patient, being the programs individual. The client is made aware of the importance of early diagnosis in this age group and that the prescription of an appropriate treatment helps prolonging the life of the dog, with better quality (Pereira, 2012a; Gioso et al. 2001b).

The customer is informed that the prevention and control of diseases, such as heart disease and silent nephropathy, is the best option for the pet, while minimizing costs (Pereira, 2012a). A check-up is the initial part of the program, with the inclusion of physical and laboratory examinations and drugs prescription, if necessary. After that, the program is individually adequate for each individual patient, as well as routine exams and the interval between medical appointments (Pereira, 2012a). The longer the dog live, the longer he will be served by the Clinic and the owner will be a satisfied customer. In today's competitive market, differentiated programs contribute greatly to customer loyalty and business growth (Pereira, 2012a).

Geriatrics, dentistry, nutritional counseling, pelage care are the most timely clients activation, however, there are many other services and medical specialties. Since nobody knows everything, the recommendation is to make partnerships with other colleagues, outsource various medical specialties, keeping your customers satisfied and present in your clinic or hospital (Pereira, 2007). An annual Marketing schedule can be done, choosing the month of geriatrics, with offers for the geriatric profiles. Each month, there may be a type of services discount, and thus, with defined calendar, it is possible to prepare educational materials and train employees outlining the details of monthly activities (Pereira, 2007). The creation of this calendar attracts new customers and strengthen ties with existing ones.

Some services, today, are popular among pet owners and can improve and increase the business profitability, as suggested by SEBRAE (Brazilian Service of Support for Micro and Small Enterprises).

Happy Day: the owner leaves his dog for a whole day, so he can play with other dogs. The dog has ten hours of fun (from 8 a.m. to 6 p.m.);

Dogs daycare or gym: in the gym, the animals performs activities to improve their behavior, causing them to spend energy, ensuring the balance of the animal and reducing hyperactivity, aggression and anxiety problems. Daycare is open Monday to Friday from 7 a.m. to 7 p.m., and offers three meals daily, collective play, walks, relaxing, dry baths and manners lessons, involving basic training;

Therapeutic treatment: in the form of massage for dogs, it helps treating spine problems, strengthens muscles, balances the immune system and promotes tranquility. Various massage techniques may be used such as, shiatsu, yoga and even lymphatic drainage. Half an hour massage costs around 45.00 reais. The most important is that the cost for the provision of this service may be zero, if the entrepreneur makes partnerships or use the space of his own clinic;

Care at home: services such as bathing and grooming can be made at the customer's home; in addition, accessories, toys and food can be taken to be sold;

Modern health care: Veterinary medicine is getting closer to the human, then, there is a variety of equipment, medicines and modern treatments to look after the health of the pet. Among the devices, there is one that maps the animal bloodstream and another that evaluates spine and brain lesions and can detect tumors.

5. PROTECTIVE SHELTERS OR DOGS PROTECTION ASSOCIATIONS TREATMENT

For this work, it is essential to have honesty and clarity from the beginning, in the form of an open conversation, with explanation of the conditions of service, discounts and payment methods (Biele, 2012).

With the care of dogs in shelters or associations, you can raise awareness of local community about specific Clinic and its social responsibility, which can attract customers (Wasche, 2008). When combined the interests of public and organizations, a strong relationship between Marketing and Public

Relations is established, contributing to the results to be achieved, and if there is a profit, it should be resorted to Integrated Marketing Communications (Pitombo and Pizzinatto, 2005).

6. EUTHANASIA

If there is the need for such a procedure, it is best to talk with the owner carefully, expose the real condition of euthanasia indication, with an ethical and impartial attitude. The professional must respect the necessary time for the owner to assimilate, reflect and decide in relation to this act. After the decision, the veterinarian must comfort the owner and obtain a documented consent of the procedure (Stewart, 2008).

There isn't something more disturbing to customers, while waiting for an appointment, than to witness the departure of an owner in tears, taking a black trash bag with him, carrying a euthanized patient. This episode negatively impacts on the company image (Stewart, 2008).

In the indication of procedure and its implementation, the Vet and his team must act with respect, stating to the owner the indication of euthanasia at the ambulatory and not at the reception. Euthanasia indication or signature of the term, should not be held in the reception with other customers witnessing the situation. The veterinarian and his crew must show compassion, solidarity and comfort the owner (Stewart, 2008).

7. OWNER'S GUIDANCE ON THE ACQUISITION OF NEW DOG

Some points should be considered when acquiring a new dog. Thus, it is necessary to know whether the customer had a pet before, if he already has pets, his lifestyle, if he has the physical space at home, availability of time and needed financial investment.

When the owner already has animals, we must clarify that the new pet will get along with each other, and this can result in ruffles. In the case of the presence of children under the age of six, one must consider that these require special attention, and that, at this age, children are more active and dogs that live with these children may become also more active and aggressive.

The customer can adopt a dog in shelters or associations, or purchase a specific dog in kennels. In the latter case, the vet should have suggestions of possible purchasing contacts and make them available to the client; the services provided in his Clinic, such as vaccines, grooming facilities and options for available diagnosis and medical specialties must be also presented (Lofflin, 2007).

By the dialogue with the client, its needed to tell him that prevention is very important and it costs 30 - 40 % less than any prompt treatment, in addition to avoiding animal suffering (Pereira, 2009b).

When adopting a dog, the owner must be aware of the commitment that he is taking, because some people, after feeling the weight of this responsibility, leave the animal because it does not fit their lifestyle (Lofflin, 2007).

8. CUSTOMER SATISFACTION

For greater satisfaction of the dog owner, the receptionist, nurses and veterinarians must work at the same pace (so they're always accessible to the customer, responding to their inquiries and phone calls). Clients must be treated with education, compassion, good mood and patience. The team must demonstrate care and concern for pet, for their stories and listen to the concerns and needs of the owner (Stewart, 2008). In difficult customer satisfaction situations, the professional should ask himself:

o   Why is the customer unhappy?
o   What should I do for the customer to feel happy in the situation?
o   Is this something we could have done for the client? Why or why not?
o   If not, how could we have communicated better or helped the customer to find what he needed elsewhere?

If customers are leaving a particular clinic, you need to know why. Thus, a letter may be sent to the client, asking the reason for the disconnection, or providing a special offer for the establishment served services, trying to bring him back (Stewart, 2008).

The act of purchasing does not arise magically. It starts with a motivation, that will lead to a need, which, in turn, awakens a desire. It then gives rise to preferences for certain specific ways to meet the initial motivation, being these preferences directly related to self, i.e., the client will tend to choose a product or service that matches the concept that he has or would like to have of himself. However, in the opposite direction to motivation, there are the brakes which are the awareness of risk, explicitly related to the product or service (Karsaklian, 2000). So it is not advisable to offer anything to the client but focus the attention into the patient and show to the client the benefits that a product or service will bring to his pet, because this is the moment of truth.

9. Owners Awareness on Monitoring of Chronic Diseases and Their Costs

Cardiovascular, joint, endocrine, neoplastic and chronic kidney diseases require a frequent and prolonged doctor monitoring, which brings costs. Accordingly, the veterinarian should be clear and properly explain to the owner the diagnosis, prognosis, the tests that should be performed during monitoring, the treatment options, diet management and risk conditions and costs that these diseases entail.

It is recommended to make an informative brochure about every disease, which must be delivered to the owner, according to the disease in question.

These folders give the owner a better understanding of the commitment and personal responsibility, financial and time availability that he would have to show to control the disease. The owner properly verbally informed and in possession of the folder, when returning home will carefully read it, and realize what effect the disease can have on the health of his pet. In this condition, the owner better accepts the condition of monitoring and treatment, and understands its costs.

If the veterinarian do not pay the proper attention and do not explain clearly to the owner the disease which his animal carries, the same will search about the subject over the internet, books and with other professionals (Keen, 2013).

10. Encouraging Self-Esteem of Your Team

The work team should be organized, its functions should be distributed, being each one in charge of certain tasks. This reduces stress and allows the attending of other customers to occur efficiently (Barforoush, 2012).

All team members are important and should be valued. One difficulty can be challenging and/or an opportunity in the form of a continuous learning. The employee should be encouraged in their professional and personal growth, getting suggestions and if it is the case, financial assistance in attending improvement courses (Barforoush, 2012).

The development of people in organizations is broader than imagined, including, besides to training, other tools, such as coaching, social responsibility activities, career management, trainee programs, job rotation, performance management and feedback (Vilas Good and Andrade, 2009).

Pizzinatto and Soares (2004) found that actions to internal employees reflect externally to the companies.

As tools, if used in an integrated and complementary way, they denote excellence in providing services and in the care and treatment of their patients and clients.

REFERENCES

Barforoush, A. 2012. 8 ways to be 'the world's greatest veterinarian'. *Veterinary Economics*, Dec, 1, 2012. DVM 396 newsmagazine. On-line: http://veterinarybusiness.dvm360.com/vetec/ Veterinary +business/8-ways-to-be-the-worlds-greatest-veterinarian/ArticleStandard/Article/detail/797475?contextCategoryId=36289.

Becker, M. 2013. Tell the pet's owner how you really feel. *Veterinary Economics,* Feb, 1, 2013. DVM 396 newsmagazine. On-line: http://veterinarybusiness.dvm360.com/vetec/Veterinary+business/Tell-the-pets-owner-how-you-really-feel/ArticleStandard/Article/detail/804056.

Biele, H. 2012. 3 ways your veterinary practice can avoid problems with animal shelters, rescue groups. Do good for pets and pet owners, and maintain your bottom line with smart relationships with rescue groups and shelters. *Veterinary Economics* Nov, 28, 2012. DVM 396 newsmagazine. On-line: http://veterinarybusiness.dvm360.com/vetec/Veterinary+business/3-ways-your-veterinary-practice-can-avoid-problems/ArticleStandard/Article/detail/797771.

Felsted, K.; Brakke, J. 2013. You're not bringing in new veterinary clients? If you can't remember the last time you introduced yourself to a client, that's a problem. *Veterinary Economics*, Feb, 1, 2013. DVM 396 newsmagazine. On-line: http://veterinarybusiness.dvm360.com/vetec/Veterinary+business

/Youre-not-bringing-in-new-veterinaryclients/ArticleStandard/Article/detail/804060?contextCategoryId=45467.

Giosso, M. A.; Kornfeld, S.; Rodrigues, M. G. J. 2011a. Seja responsável pela experiência de seus clientes – parte 1A: Desde o telefonema até a sala de exame. *Clínica Veterinária*, ano XVI, N°92, 2011a, pp. 110-111, 1413571-X.

Giosso, M. A.; Kornfeld, S.; Rodrigues, M. G. J. 2011b. Seja responsável pela experiência de seus clientes – parte 1B. *Clínica Veterinária*, ano XVI, N° 93, 2011b, pp. 108,1413571-X.

Giosso, M. A.; Kornfeld., S.; Rodrigues, M. G. J. 2011c. Seja responsável pela experiência de seus clientes – parte 2. *Clínica Veterinária*, ano XVI, N° 95, 2011c, pp. 110, 1413571-X.

Karsaklian, E. Comportamento do consumidor. São Paulo: Atlas, 2000.

Keen, J. 2013. Prep clients for long-term pet care. *Veterinary Economics*, March, 1, 2013. DVM 396 newsmagazine. On-line: http://veterinary-business.dvm360.com/vetec/Veterinary+business/Prep-clients-for-long-term-pet-care/ArticleStandard/Article/detail/   806836?   Context CategoryId=8804andref=25.

Kotler, P. Administração de marketing. 10. ed. São Paulo: Prentice Hall, 2000.

Las Casas, A. L. 2008. Administração de marketing: conceitos, planejamento e aplicações à realidade brasileira. São Paulo: Atlas, 2008.

Lobato, S. 2010. É preciso ser mega? *Clínica Veterinária*, ano XV, N 85, 2010, pp. 102, 1413571-X.

Lofflin, J. 2007. Assist owners in selecting the best pets for their lifestyles. *Veterinary Medicine*, Oct, 1, 2007. DVM 396 newsmagazine. On-line: http://veterinarymedicine.dvm360.com/vetmed/article/articleDetail.jsp?id=463624.

Maccarthy, E. J. Basic Marketing: a managerial approach. 12. ed. Homewood, II: Irwin, 1996.

Materni, C.; Tumblin, D. 2013. 5 ways to #market your @vetpractice on social media. Say goodbye to old-school advertisements and hello to online engagement with clients. It's going to be quick and painless—we promise. *Veterinary Economics* Jan., 1, 2013. DVM 396 newsmagazine. On-line: http://veterinarybusiness.dvm360.com/vetec/Veterinary+business/5-ways-to-market-your-vetpractice-on-social-media/ArticleStandard/Article/detail/801058?contextCategoryId=45467.

Pereira, M. S. 2005. Especialidades Médicas (II). *Clínica de Marketing – Nosso Clínico,* ano 8, N° 45, 2005, pp. 189-192, 1808-7191.

Pereira, M. S. 2007. Tempo ocioso é custo. *Clínica de Marketing – Nosso Clínico*, ano 10, N° 60, 2007, pp. 249-252, 1808-7191.

Pereira, M. S. 2008. Entender para atender. Faça com que os proprietários conheçam as necessidades médicas do animal de estimação. *Clínica de Marketing – Nosso Clínico*, ano 11, N° 63, 2008, pp. 261-264, 1808-7191.

Pereira, M. S. 2009a. Ambiente competitivo exige excelentes e economicamente atentos. *Clínica de Marketing – Nosso Clínico*, ano 12, N 71, 2009a, pp. 293-296, 1808-7191.

Pereira, M. S. 2009b. Clínica boca a boca. *Clínica de Marketing – Nosso Clínico*, ano 12, N° 72, 2009b, pp. 297-300, 1808-7191.

Pereira, M. S. 2010a. Ninguém compra o que desconhece. *Clínica de Marketing – Nosso Clínico*, ano 13, N° 75, 2010a, pp. 309-312, 2010ª, 1808-7191.

Pereira, M. S. 2010b. Não é apenas a quantidade e sim a qualidade da comunicação que fará a diferença. *Clínica de Marketing – Nosso Clínico*, ano 13, N° 76, 2010b, pp. 313-316, 1808-7191.

Pereira, M. S. 2010c. A administração da clientela, quando feita adequadamente, resultará em importante impacto na receita. *Clínica de Marketing – Nosso Clínico*, ano 13, N° 77, 2010c, pp. 317-320, 1808-7191.

Pereira, M. S. 2011. Clientes inativos. *Clínica de Marketing - Nosso Clínico*, ano 14, N° 84, 2011, pp. 345-348, 1808-7191.

Pereira, M. S. 2012a. Programa de geriatria animal. *Clínica de Marketing – Nosso Clínico*, ano 15, N° 85, 2012a, pp. 349-352, 1808-7191.

Pereira, M. S. 2012b. O mercado das clínicas veterinárias. *Clínica de Marketing – Nosso Clínico*, ano 15, N° 86, 2012b, pp. 353-356, 1808-7191.

Pereira, M. S. 2012c. O Marketing deve criar e manter uma relação contínua com o cliente após a primeira compra. *Clínica de Marketing – Nosso Clínico*, ano 15, N° 89, 2012c, pp. 365-368, 1808-7191.

Pinheiro, D.; Gullo, J. 2009. Comunicação Integrada de Marketing: gestão dos elementos de comunicação e suporte às estratégias de marketing e negócios da empresa. 3. ed. São Paulo: Atlas, 2009.

Pitombo, T. D. T.; Pizzinatto, N. K. 2005. Foco no cliente: as relações públicas. In: PIZZINATTO, Nadia Kassouf. (Org.). Marketing: focado na cadeia de clientes. São Paulo: Atlas, 2005.

Salvador, M. 2007. Integrando clínicas veterinárias e pet shops com a internet. *Clínica de Marketing – Nosso Clínico*, ano 10, N° 59, 2007, pp. 245-248, 1808-7191.

Schaible, J. 2012. Why veterinarians shouldn't be allowed to blog. Blogging may be the latest trend, but evidence suggests that veterinarians should share their knowledge in other ways. *Veterinary Economics* Oct., 1, 2012. DVM 396 newsmagazine. On-line: http://veterinarybusiness.dvm360.com /vetec/Veterinary+business/Why-veterinarians-shouldnt-be-allowed-to-blog/ArticleStandard/Article/detail/ 792718.

Soares, J. A; Pizzinato, NK. 2004. O endomarketing como estratégia empresarial de transformação do ambiente de trabalho: estudo de casos. XXXIX Asamblea Anual del Consejo Latinoamericano de Escuelas de Administración (CLADEA), Pontifícia Universidad Católica Madre y Maestra, Puerto Plata, República Dominicana, 2004.

Stewart, P. 2008. Why clients leave. Five pet owners tell us why they ditched their veterinarians. Learn from their experiences—then use these tips and tools to avoid client care mistakes. *Veterinary Economics* Jun, 1, 2008. DVM 396 newsmagazine. On-line: http://veterinarybusiness.dvm360.com /vetec/article/articleDetail.jsp?id=520553andsk=anddate=andpageID=3.

Tudor, R. 2013. Good deeds are your new Yellow Pages. Looking for your next big Marketing idea? Skip the old-school approach and promote the good stuff you and your veterinary staff are doing instead. *Veterinary Economics* Feb, 1, 2013. DVM 396 newsmagazine. On-line: http://veterinarybusiness.dvm360.com/vetec/Veterinary+business/Good-deeds-are-your-new-Yellow-Pages/ArticleStandard/Article/detail/804065? contextCategoryId=45467.

Tracey, S. 2011. 9 steps to a perfect veterinary appointment: Tweak your communication and practice protocols for big dividends in client compliance and service. *Veterinary Economics*, May, 1, 2011. DVM 396 newsmagazine. On-line: http://veterinarybusiness.dvm360.com/ vetec/article/articleDetail.jsp?id=710159.

Vilas Boas, A. A.; Andrade ROB. Gestão estratégica de pessoas. Rio de Janeiro: Elsevier, 2009.

Wasche, L. 2008. A 5-step plan for helping shelters and rescue groups. *Veterinary Economics* May, 7, 2008. DVM 396 newsmagazine. On-line: http://veterinarybusiness.dvm360.com/vetec/Veterinary   +business/A-5-step-plan-for-helping-shelters-and-rescue-grou/Article Standard/Article/detail/515237.

Whiteley, R. C. A Empresa totalmente voltada para o cliente: do planejamento à ação. 18. ed. Rio de Janeiro: Campus, 1992.

In: Dogs
Editors: K. M. Cohen and L. R. Diaz

ISBN: 978-1-62808-530-3
© 2013 Nova Science Publishers, Inc.

Chapter 4

RESPONSIBLE OWNERSHIP AND BEHAVIOR

Lucas Vinicius Shigaki de Matos[1],
Julia Cestari Pierucci[1], Weslen Fabricio Pires Teixeira[1],
Alvimar José da Costa[1], Milena Araúz Viol[2],
Monally Conceição Costa de Aquino[2],
Willian Marinho Dourado Coelho[3]
and Katia Denise Saraiva Bresciani[1,2]*

[1]Faculdade de Ciências Agrárias e Veterinárias, UNESP,
Jaboticabal, São Paulo, Brazil;
[2]Faculdade de Medicina Veterinária de Araçatuba, UNESP,
Araçatuba, São Paulo, Brazil;
[3]Faculdade de Ciências Agrárias de Andradina – FCAA,
Andradina, São Paulo, Brazil

ABSTRACT

The responsible ownership has been increasingly broached in the modern
society, mainly due to the growth of emotional bond between owners and
their pets. The lack of responsible ownership can lead to many problems
for society, including the overgrowth of animal population, contributing
to the increase of stray animals and consequently greater risk of zoonosis.
In addition, the lack of responsible ownership can lead to an emergence

* bresciani@fmva.unesp.br.

of animal behavioral disorders due to the owner's carelessness. Among these disorders behavioral changes such as aggression, depression, decreased socialization with other animals and self-mutilation habits can be observed. In this context, the main issue allied to possible solutions will be broached for the subject.

INTRODUCTION

The dog has its figure printed in art since prehistory, through paintings that has no artistic intention, but only to register events and culture.

It's represented in several ways, whether with a symbolic purpose as in Van Eyck's artwork "Esponsais dos Arnolfini" as reference of loyalty, or merely like a family pet (Serafim, 2012).

By the eighteenth century, Kant suggested the "principle of autonomy", whereby man would have the rights of universal ownership, acting based on their own values. Therefore, according to this author, since the animals don't have values that correspond to human race, they would lose their individual freedom and be available to man's will.

This principle was reinforced by Hegel which, at the beginning of the next century, described ethics as a result of self-conscious human involvement, not applying to animals since they are unaware (Petroianu, 1996).

The inseparable human's instincts are classified into two groups: the erotic and the destructive (or the death one). The first has a constructive, positive and preservation nature. The second desires the aggression and destruction that leads to annihilation, both himself and others. This last one is explained by psychological basis, which elucidate the several ways that human are capable of executing the greatest atrocities and cruelties to animals, especially when there is no moral censorship in society that represses this hostile behavior (Freud, 1996).

In 2003 the definition of responsible ownership was brought up at the First Meeting of Latin-American Specialists in Responsible Pet Ownership and Control of Canine Populations, performed by the Pan American Health Organization PAHO/WHO, along with (World Society for the Protection of Animals) (Souza, 2003). This definition consist in the "condition in which the guardian of a companion animal accepts and assumes a number of duties focused on taking care of the physical, psychological and environmental needs of his pet, as well as prevent risks (aggression potential, disease transmission or harm others) that his pet can cause to the community or the environment, as interpreted and supported by the current legislation".

Nowadays cats and dogs purchased great importance in the family setting (Antunes, 2011). What was a plain animal with its guard and company function, today is considered by many as a home member, which has room, blanket such as several toys and therefore being part of the family's monthly budget.

According to Matos et al. (2012), there are many people that are willing to do what it takes to make their animal healthy and happy. However, there are some exceptions of owners that insists in mistreating their pets, not exerting a responsible ownership, leading to a severe social issue.

The responsible domestic animal ownership's issue is one of the most urgent legal arrangements of Environmental Law, behold the increasing demand established in societies due do the progressive urban growth which has overtaken collective habits among individuals that, isolated in their homes, have provided strong affectional bonds with some species, such as cats and dogs, turning them into real family members (Santana and Oliveira, 2004).

The Universal Declaration of Animal Rights infers that the concept of responsible ownership implies human conduct of giving animals respect, not submitting them to mistreatment, abuse, nor causing their extermination (Santana and Oliveira, 2004). Thus, attitudes such as vet visits, population control, deworming, immunization, restriction of street access and appropriated nutrition of animals are the owner's responsibility (Matos et al., 2012).

Irresponsible owners often abandon their mascot in the streets, that aren't castrated in most times, leading to increase of stray animals population and their consecutive generation, equally unwanted (Magalhães, 2008; Silva et al., 2009).

According to Matos et al., (2012), this event can propitiate health human injury, through major risk of appearance and transmission of zoonosis, or even increase traffic accidents rate.

In everyday life, there is much arbitrariness committed by human that annihilate the dignity of these extremely dependent and defenseless creatures, by promoting all kind of abuse, mistreatment and cruelty, or else submit them for training be violent and therefore turn them into "weapons" (Santana and Oliveira, 2004).

Animal welfare can be defined as the way that the animal relates to its environment. This is a wide definition, once the animal owns mechanisms that are adapted to natural life which can or not be useful for the conditions that they are submitted in domestic life (Broom and Johnson,1993).

An essential criterion for the definition of animal welfare is that it should refer to the individual characteristics of each animal, not to something offered to it by human. The animal's wellness may improve as a result of something provided to it, although what is given is not, by itself, wellness (Broom and Molento, 2004).

The term "well-being" can be used for people, wild animals, animals captive in productive farms, zoos, on experiments or in homes (Broom and Molento, 2004).

Illness, injuries, starvation, social interactions, housing conditions, inadequate treatment, handling, transport, laboratory procedures, varied mutilations, veterinary treatment or genetic alterations through conventional genetic selection or genetic engineering are all elements that directly affect the wellbeing (Broom and Molento, 2004).

By recognizing nutritional, environmental, health, behavioral and mental domains of welfare, it is possible to assure good welfare as existing when an animal's needs in these interacting domains are largely being met, which are more elucidated below (Mellor et al.,2009).

Compromised nutrition includes inadequate fluid or food intake, or even dietary nutrient imbalances (deficiency or excess). These factors can lead to disorders such as hunger, thirst, weakness and debility.

The environmental influence on animal welfare can be verified in both outdoor and indoor. Referring to outdoor atmosphere, exposure to extreme weather such as cold or hot can lead to, respectively, hypothermic or hyperthermic disorders. Yet in indoor surroundings stressing and injuring situations associated to physical structures such as the floor material in constant contact with the animal, can bring pain from bruises, join injury, persistent discomfort, and so on (Mellor et al., 2009).

Compromised health may occur in response to traumatic injury. Thereby a physically injured animal (tumors, fractures and lesions) already has its welfare reduced or is at risk sometime in the future of having its welfare reduced.

Also disease agents or toxins, genetic disorders may lead to verity number of unpleasant experiences including breathlessness, nausea, pain, distress, fear or anxiety.

Behavioral changes due to human behavior create negative effects on animal welfare. According to Mellor et al., (2009) dogs with behavioral problems are less likely to be interact by their owner, thereby are less exercised and walked in public.

The lack of activities may lead to a wide number of problems which can me associated to animal's health (Biourges, 1997; Morgante, 1999). According to Lewis et al., (1987) and Camps, (1992), there is a direct correlation between dog and its owner's overweight.

Obesity due to excessive calorie ingestion and lack of exercises (Case et al.,1998) is one of the most important elements that can lead to problems such as Diabetes mellitus (Ferreira and Carvalho, 2002), mobility difficulties (Case et al.,1998), cardiac and respiratory insufficiencies, (Wilkinson and Moonney, 1990), skin, reproductive and neoplastic diseases (Ward, 1984; Alenza et al., 2000).

Distraction and physical contact with its owner can be achieved through toys, paper bags or boxes and by entertainment places such as parks, animals playgrounds and so on. Another option, commonly adopted, is the use of toys stuffed with foods that stimulate the animals interest consequently most exercise (Genaro, 2004).

Another common problem related to animal welfare refers to its abandonment, which leads to unrestrained procreation, implying in overpopulation of stray animals and, on the other hand, increasing zoonotic potential diseases dissemination (Matos et al., 2012).

The species distinction and the acquaintance of each one are essential, since every animal has its own attribute and characteristic. To provide the animals' proper living, even inside tightly spaced homes, the species' evolutionary history must be carefully considered.

Cats unlike dogs are animals that express their behavior seamlessly. In general they respond to poor environmental conditions with indifference and also reducing certain natural behaviors such as self-cleaning and displacement.

Every breed requires special conditions to survive, and these needs have to be considered at the moment of choosing a pet by a beginner owner.

Some actions indicate lack of awareness and responsible ownership resulting in problems, such as owners who feed their dogs with homemade food, which, if unbalanced, may lead to nutritional deficiencies and compromised immune system of these animals (Motta, 2009).

Allow the animal's street access increases the probability of it being run over by vehicles, or cause traffic accidents, obtain injury as a result of fights with other animals, or even bite people or be affected by bacterial diseases, or parasitic infestation or infection (Santana and Marques, 2002; Souza, 2003; Santana and Oliveira, 2004).

Bathe the animal every seven days in summer and every two weeks in winter is considered ideal, depending on whether the dog is kept in yard or apartment. However this daily procedure is contraindicated, except in cases with scabies and seborrhea, since it can remove the animal's natural oil, letting it more vulnerable to infestation by parasites and dermatitis.

Rarely or never take your pet to be examined by a veterinarian is a worrying fact for animal and human health (Matos et al., 2012).

In the preventive aspect of the main viral and parasitic diseases, there are animals that are found unvaccinated or not dewormed, which are wrong conducts practiced by owners, since prophylactic attitudes are indispensable for the animal's immunity against several diseases, even the ones with zoonotic potential (Who, 1990, Who, 2005).

Not adopting procedures such as animal castration or the lack of use of some type of contraceptive treatment provides an evidence of misconduct in animal population control (Schoendorfer, 2001; Magalhães, 2008).

ANIMAL BEHAVIOR

Animal behavior can be defined as the association of several activities that an animal performs, such as locomotion, communication and reproduction, also including reactions to different stimulus both an environmental and social. These reactions may be influenced by genetic factors (aggressiveness or docile) or by environmental factors (living and learning with other animals and man), and therefore may affect the animal's behavior in different ways (Braastad and Bakken, 2002).

Studies have shown that behavior is a quantitative genetic trait and results from the interaction of numerous genes, allowing the recognition of distinct phenotypes.

The classical ethological model supports the importance of animals to be allowed to exhibit species typical instincts (earned during its evolutionary adaptation). According to Weary and Fraser (2002), despite animal domestication, these instinctive characteristics are still expressed with humans, the same way when in contact with other animals, such as the protection behavior to their family members and dominance relationship.

By means of ethology it is possible to understand many behaviors that, today, have no apparent function. Only multidisciplinary research allows an actual understanding of animal behavior inside our homes (Rossi, 2008).

Every animal's attitude has a meaning and its understanding is necessary for human to learn how to act in several situations. Any owner's wrong decision or conduct can lead to several consequences, similar to traumas, for the rest of its life.

For example, the dog needs company; it's just as dependent as the wolves, which need each other to survive. The solitary specimen could not hunt large animals and confronted greater difficulties to protect themselves compared to the wolf that lived in groups. Therefore, being alone may lead to death. Thus the specimens that predominated in reproduction were, in major part, company dependent (Rossi, 2008).

Many current animals' behaviors can be explained by the behavior of their ancestral. Therefore some dogs' actions such as bury bones or toys between pillows, despite being in an environment that there would be no risk of losing it, can be explained by their ancestral behavior, which also hid food, allowing the transfer of this instinct thru generations.

While other less cautious specimens were affected by hunger, the most nourished could defend themselves and reproduce better. Therefore, they left more descendents and spread their own genetic characteristic. By this event the naturalist Charles Darwin created the evolution theory.

Inherited behaviors are not constant, however, some attitudes are difficult to change, but others are simple. When the instinctive behavior leads to problems, it is important trying to change the way the animal is programmed to understand things. And if the animal presents some extremely dangerous attitude, behavioral techniques can be applied in an attempt to inhibit it. As an example, a common situation of a child, who is mistakenly associated with a prey and attacked by a dog. In the situation shown, we may not be justifying. A skilled professional should be looked for. Since these attitudes are present in its instincts, hitting the animal will not solve the issue. The owner need to say "no" with plenty of patience and show the animal what is the issue situation which may not repeat (Rossi, 2008).

MAJOR BEHAVIORAL DISTURBANCES RELATED TO RESPONSIBLE OWNERSHIP

In many situations, animals can be used to improve human life, and this strong bond is considered benefic for human. In contrast, the animal's illness or its misbehavior may lead to negative effects on the owner.

The behavior of animals living in close contact with humans can be influenced by human attitudes. Therefore, animals can easily acquire inappropriate behaviors such as barking. Furthermore, it is believed that human habits can encourage or reinforce timidity, anxiety and aggressive behavior in their pets (Wong et al., 1999).

Stereotypy is considered a kind of abnormal behavior and it can be described as repeated movements, unvarying, with apparently no obvious goal or function and high frequency.

This behavioral disorder can be easily diagnosed, although can also be less obvious. Usually this abnormal habit occupies a considerable part of the time during which the animal is awake, and it's considered a significant animal welfare indicator, being closely related and influenced by the environment that surrounds the animal. Thus a restrictive environment, with poor stimulus may deprive the animal to express natural attitudes, leading to stereotypies development.

Stress is therefore another important behavioral abnormality, and, according to Broom and Johnson (1993), it is considered an important bond between environment, behavioral disorders and disease.

The abnormality above is a mainly behavioral phenomenon and considered a valid indicator of well-being due to its simple detection and by representing a reflect of the animal's reaction to a certain situation (Jensen and Toates, 1993).

Defining well-being patterns for companion animals is a difficult task because they are often treated and interpreted as human beings, inserted most of the times into the family context. This type of treatment commonly causes behavioral disorders in these animals (Overall, 1997), and the Separation Anxiety Syndrome in Animals (SASA) is one of the most observed.

The Separation Anxiety Syndrome in Animals (SASA) is defined by a set of undesirable behavior exhibited by the animals when they are left alone or when they are physically separated from the bond or affection figure, which can be represented by a human or another animal (Soares et al., 2010).

The behaviors that most often characterize the SASA are: excessive vocalization (howling, whining or barking excessively), destructive behavior (chew or scratch personal objects that may connect it to the bond or affective figure), urination and defecation in inappropriate places, or specific places that are linked to the affective figure. (Mccrave, 1991; Overall, 1997; King et al., 2000; Beaver, 2001; Appleby and Pluijmakers, 2003; Schwartz, 2003; Landsberg et al., 2004).

However, there are other behaviors considered manifestations of this syndrome, such as vomiting, sialorrhea and depression (Mccrave, 1991; Landsberg et al., 2004), including comorbidity with compulsive disorders (Mccrave, 1991; Overall and Dunham, 2002; Landsberg et al. 2004).

The depressive behaviors expressed during SASA are characterized by total inactivity dog, that is, the animal stops urinating, defecating or even eating, and there are reports does not urinate, defecate there, do not eat, and generally, according to reports, the animal express drowsiness throughout time alone.

Anxiety-related disorders are frequently reported as the most common class of problems in pets (Overall, 1997)

According to Mellor et al. (2009), chronically anxious dogs and cats are common, and suggest management deficiency of these animals.

Among the behavioral problems in older pets, separation anxiety, aggression to people and to other animals, excessive vocalization, inadequate control of physiological needs, noise phobias, compulsion and sleeping disorders are mostly common (Landsberg and Araujo, 2005; Freitas et al., 2006).

A frequent mistake occurs when assigning the animals breed with behavioral problems by chance diagnosed. An example is the Siberian Husky breed, which was created to travel for days without stopping, a natural characteristic of these dogs. Its genes have the capacity to run across longer distances. This innate ability, however without any sufficient training, will lead the animal to frustration and present, for example, the same symptoms as any dog like a dachshund or pit-bull frustrated would present. Nervousness, fear, aggression, tension and territorial behavior: all these difficulties and disorders appears when an animal is frustrated. Therefore it is a mistake to analyze any breed on its suspicious or uncertain condition (Millán and Jo Peltier, 2007; Millán and Jo Peltier, 2011).

Death or withdrawal of close animals, or even the introduction of a new pet in the same rearing environment are issues that lead to social disorders. (Houpt and Beaver, 1981).

The animal's living conditions (management practices and environment) may contribute to the development of aggression. According Ciampo et al. (2000) elderly people, children, acquaintances, strangers, owners or visitors may be the target of these reactions, and are attacked when the animal feels threatened with no ways to defend itself.

Such aggressive behavior in general, begins in the dog's early years, when there weren't any limits enforced, and it recognized itself as a leader, or as dominant (Rocha, 2003).

According to Rossi (1999), to minimize those aggression reactions, the animals must be raised in appropriated atmosphere for the specie, providing adequate food, in the recommended frequency and amount, as well as abundance of clean and filtered water, distraction toys, among others.

The lack of an environment that provides the puppies contact with other people and other animals is also an important element for developing abnormal behaviors (Rossi, 1999; Viaro, 2004). According to Rossi (1999), punishment or excessive "pampering", as well as isolation from unknown people and inadequate housing can lead to development of an aggressive animal.

Genetic elements are also related to aggressive disorders, although owners themselves who doesn't provide satisfactory living conditions or correct management of their pets are the most often responsible for the appearance of this unwanted behavior in animals (Beaver, 2001).

According to Rocha (2003), every animal species is composed by their own breeds that are, in turn, more aggressive than others. However, aggressiveness may not be attributed only to breed component, but shall be considered that the relationship between man and his pet should be harmonious and pleasant, and that abuse and aggression reflect a failure of this harmony (Beaver, 2001).

Behavioral compromise can result from severe space restrictions, or overcrowding, or even the interaction of multiple factors, which can lead to an inhibition the animal's natural behavior, such as foraging, hunting, play, exploration and positive social interaction.

The animal has to be able to recognize its owner as a member of its family to maintain its psychological balance. For this purpose, the owner must recognize the best way to interact with his pet. Otherwise, a destructive and aggressive behavioral disorder can be observed (Resende et al., 2006).

CONCLUSION

Recent studies suggest that the interaction between man and pets can improve human's physical, mental and emotional health, providing physiological, psychological and social effects.

Psychological effects: humor improvement as well as reducing depression, stress, anxiety, feelings of inferiority and loneliness.

Physiological effects: lower blood pressure and heart rate, low cholesterol levels, higher survival rates after a heart attack, higher self-esteem, greater life expectancy, and motivation for healthy activities.

Social Effects: socialization of criminals, elderly, disabled and mentally people; improvement in children's learning and socialization. Anyway, possession of a dog may mediate the relationship between physical activity and disease outcomes because it encourages physical activity.

Therefore, owning a dog is an experience that may mediate the bond between physical activity, behavioral characteristics and the course of a disease, due to its beneficial intimate association to psychological, physiological and social effects.

The development of the relationship between human and companion animal is at the essence of an important society behavioral change, which began to cultivate several habits such as: fewer children among with more resources in general, give the pet the status of a family member who happens to inhabit most indoors than outdoors. The pet earns its space and becomes part of the family's budget, and is taken care since birth until death.

On the other hand, in a part of the population it is possible to observe a lack of information by the owners, suggesting the importance of an implantation of educational campaigns and public awareness programs focusing on responsible ownership and control of zoonosis, that way decreasing the risk of diseases dissemination.

Therefore, university extension projects and public management are clearly necessary, targeted in society awareness, by mean of regular campaigns on population behalf, with current information about correct attitudes of welfare and handling of pets. Such changes in the conduct of both the owners and government agencies will incur reduction in animals abandonment and consequently decrease zoonosis transmission, improving the quality of life of our "fellows".

REFERENCES

Alenza, MDP; Pena,L; Del Castillo, N; Nieto, AI. Factors influencing the incidence and prognosis of canine mammary tumours. *Journal Small Animal Practic, Oxford*, 2000.v. 41, n. 10, pp. 478.

Antunes, MR. Zoonoses parasitárias. *Revista Brasileira Medicina*. 2001, *v.* 58, pp. 661-662.

Appleby, D; Pluijmakers, J. Separation anxiety in dogs: the function of homeostasis in its development and treatment. *Veterinary Clinics of North America: Small Animal Practice*, 2003, v.33, n.2, pp.321-344.

Bahr, SE; Morais, HA. Pessoas imunocomprometidas e animais de estimação. *Revista Clínica Veterinária*. 2001, v.30, pp. 17-22.

Beaver, BV. Comportamento canino: um guia para veterinários. São Paulo: Roca, 2001. pp. 431.

Biourges, V. Obesidade. *Informativo Técnico e Científico,* Centro de pesquisa e desenvolvimento da Royal Canin, 1997. [On line]. Disponível : <http://linkway.com.br>. [Data de acesso: 10 mar. 2013].

Braastad, BO; Bakken, M. (2002).*The etology of the domestic animals*: *Behaviour of Dogs and Cats*. (ed. Per Jensen), pp.173- 192. Linköping, Sweden.

Broom, DM, Johnson KG. Stress and animal welfare. London: Chapman and HALL, 1993.

Broom, DM; Molento, CFM. Bem-estar animal: conceitos e questões relacionadas – Revisão. Archives of Veterinary Science, 2004, v.9, n.2, pp.1-11.

Camps, J. Manejo de la alimentación en los distintos estadios. *Medicine Veterinary*, 1992, v. 9, n. 5, pp. 321-325.

Case, LP; Carey, DP; Hirakawa, DA. Nutrição Canina e Felina. *Manual para Profissionais*, Madrid – Espanha: Harcourt Brace de España S. A., 1998, pp. 424.

Ciampo, LAD; Riccoa, RG; Almeida, CAN; Bonilhac, LRCM; Santos, TCC. "Acidentes de mordeduras de cães na infância", *Revista de Saúde Pública*, São Paulo, 2000, v. 34, n° 4, pp. 411-412.

Ferreira, F; Carvalho, AU. Manejo nutricional de cães e gatos com Diabetes Melito. *Cadernos Técnicos de Veterinária e Zootecnia da UFMG*, Belo Horizonte: Ed. FEP – MVZ, 2002, n. 37, pp. 39 - 45.

Freitas, EP; Rahal, SC; Ciani, RB. Distúrbios físicos e comportamentais em cães e gatos idosos (physical and behavioral disturbance in aging dogs and cats). *Archives of veterinary science*, 2006, v. 11, n. 3, pp. 26-30, printed in Brazil issn: 1517-784x.

Freud, S. Por quê a guerra? in Obras Completas de Sigmund Freud: edição standart brasileira; com comentários e notas de James Strachey; em colaboração com Anna Freud. Volume XXII. Trad.: Jayme Salomão. Rio de Janeiro, IMAGO, 1996. pp. 202-203.

Genaro, G. Comportamento felino :organização social e especial, comunicação intra-específica e conflitos com a vida doméstica. MEDVEP. Revista Científica Medicina Veterinária Pequenos Animiais Estimação, 2004; 2(5): pp. 61-6.

Houpt, K.A.; Beaver, B. Behavioral problems of geriatric dogs and cats. *The Veterinary Clinics of North America, Small Animal Practice*, 1981, v.11, n.4, pp. 643-652.

Jensen, P and Toates, FM. (1997) Stress as a state of motivational systems. *Applied Animal Behaviour Science*, v. 53, 145–156.

King, JN; Simpson, BS; Overall, KL; Appleby, D; Pageat, P; Ross, C; Chaurand, JP; Heath, S; Beata, C; Weiss, AB; Muller, G; Paris, T; Bataille, BG ; Parker, J; Petit, S; Wren, J. Treatment of separation anxiety in dogs with Clomipramina: results from a prospective, randomized, double-blind, placebo-controlled, parallel-group, multicenter clinical trial. *Applied Animal Behavior Science*, 2000, v.67, pp.255-275.

Landsberg, G; Hunthausen, W; Ackerman, L. *Problemas Comportamentais do Cão e do Gato*. São Paulo: Roca, 2004, pp. 492.

Lewis, LD; Morris, ML; Hand, MS. Obesity. In: __. *Small Animal Clinical Nutrition III*, Topeka: Mark Morris Institute, 1987. pp. 6-1 – 6- 39.

Magalhães, DN. Escolares como multiplicadores da informação sobre Leishmaniose Visceral no contexto familiar: elaboração e análise de modelo, 2008. *Tese* Doutorado Universidade Federal de Minas Gerais.

Matos, LVS; Teixeira, WFP; Bregadioli,T; Aquino, MCC; Viol,MA; Bresciani, KDS. Orientação sobre posse responsável em uma área endêmica para Leishmaniose Visceral Canina. *Revista Ciências Extensão*, 2012, v.8, n.3, pp.37.

Mccrave, EA. Diagnostic criteria for separation anxiety in the dog, Veterinary Clinics of North America: Small Animal Practice, 1991, v.21, pp.247-256.

Mellor, DJ; Kane, EP; Stafford, KJ. The Sciences of animal Welfare. Willey-Blackwell. 2009, pp. 212.

Millán, C; Jo Peltier, M. El encantador del perros, 2007.

Millan, C; Jo Peltier, M. O Encantador de Cães: Compreenda o melhor amigo do homem. Tradução de Coelho, Carolina Caires. 18.ed. São Paulo: Ed. Verus, 2011. pp.188.

Morgante, M. Obesità Negli Animali da Compagnia: problema emergente. *Praxis Veterinaria*, 1999, v. 20, n. 2, pp.18-22.

Motta, RR. Bom Pra Cachorro. São Paulo Gente, 2009.

Overall, KL. *Clinical behavioral medicine for small animals*. St. Louis: Mosby – Year Book, 1997. pp. 544.

Overall, KL; Dunham, A. Clinical features and outcome in dogs and cats with obsessive-compulsive disorder: 126 cases. Journal of American Veterinary Medical Association, v.221, n.10, pp.1445-1451, 2002. Disponível em: < http:// dx.doi.org/10.2460/javma.2002.221.1445>. Acesso em: 04 mar. 2013. doi: 10.2460/javma.2002.221.1445.

Petroianu A. Aspectos éticos na pesquisa em animais. *Acta Cirúrgica Brasileira*, 1996; v.11, pp.157-164.

Resende, DM; Machado, HH; Muzzi, RA; Muzzi, LA, 2006, "Manejo de Cães". Disponível em: http://www.editora.ufla.br/Boletim/pdfextensão/ bol_08.pdf. Acesso em 10 mar 2013.

Rocha, RR, 2003."Relação Homem-Animal e Agressividade Canina", pp. 32. Monografia (Graduação em Medicina Veterinária) – Faculdade de Medicina Veterinária, Universidade Federal de Uberlândia, Uberlândia.

Rossi, A, 1999, In:" Adestramento Inteligente', São Paulo: CMS, pp.260. Schwabe, C, 1984, *"Veterinary Medicine and Human Health"*, 3ª ed. Baltimore: Williams and Wilkins, pp.680.

Rossi, A. Comportamento canino - como entender, interpretar e influenciar o comportamento dos cães. On-line version ISSN 1806-9290 *Revista Brasileira de Zootecnia*. v.37 no.spe Viçosa July 2008. http://dx.doi.org/ 10.1590/S1516-35982008001300007.

Santana, LR; Marques, MR. Maus tratos e crueldade contra animais nos Centos de Controle de Zoonoses: aspectos pp. 551. *Anais* do 6º Congresso Internacional de Direito Ambiental, de 03 a 06 de junho de 2002: 10 anos da ECO 92: O Direito e o Desenvolvimento Sustentável. São Paulo: IMESP, 2002.

Santana, LR; Oliveira, TR. Guarda Responsável e dignidade dos animais 8., 2004. Disponível em:http://www.abolicionismoanimal.org.br/artigos/ guardaresponveledignidadedosanimais.pdf. Acesso em 10 mar. 2013.

Schwartz, S. Separation anxiety syndrome in dogs and cats. Journal of American Veterinary Medical Association, v.222, n.11, pp.1526-1532, 2003. Disponível em: <http:// dx.doi.org/10.2460/javma.2003.222.1526>. Acesso em: 12 mar. 2013. doi: 10.2460/javma.2003.222.1526.

Serafim, SSB. Universidade do extremo sul catarinense - Unesc curso de artes visuais bacharelado estudo iconográfico da representação dos cães na arte Criciúma, jun- 2012.

Silva, FAN; Carvalho, RL; Klein, RP; Quessada, AM. Posse responsável de cães no bairro Buenos Aires na cidade de Teresina (PI). *ARS VETERINARIA, Jaboticabal, SP, 2009, v.25, n.1, pp. 014-017.*

Schoendorfer, LMP. Interação homem-animal de estimação na cidade de São Paulo: o manejo inadequado e as consequências em saúde pública. São Paulo: Universidade de São Paulo, 2001. pp.82. *Dissertação* (Mestrado em Saúde Pública) – Faculdade de Saúde Publica, 2001.

Soares, GM; Pereira, JT; Paixão, RL. Estudo exploratório da síndrome de ansiedade de separação em cães de apartamento. *Ciência Rural*, mar, 2010, v.40, n.3, pp.548-553.

Souza, MFA. (org.). Resumo da Primeira Reunião Latino-americana de especialistas em posse responsável de animais de companhia e controle de populações caninas. *Primeira Reunião Latino-americana de especialistas em posse responsável de animais de companhia e controle de populações caninas*, de 01 a 03 de setembro de 2003. Rio de Janeiro, 2003 (Documento inédito).

Viaro, O, 2004. Manual do Educador, Criando um amigo. Centro de Controle de Zoonoses, *Gerência de Vigilância Ambiental, Secretaria Municipal da Saúde*, Prefeitura do Município de São Paulo.

Ward, A. The fat-dog problem: how to solve it. *Veterinary medicine*, 1984., pp.781-786.

Weary, D.M and Fraser, D. (2002). *The etology of the domestic animals: Social and Reproductive Behaviour*, pp.74-75, Vancouver, Canada.

Who. (1990). World Health Organization. *Guidelines for dog population management*. Geneva: WHO/WSPA

Who. (2005). World Health Organization. *Technical Report Series 913*. Geneva: WHO/WSPA.

Wilkinson, MJ; Moonney, CT. Obesity in the dog. *A monograph*. University of Glasgow, Department of Veterinary Medicine. Walthan. 1990, pp.19.

Wong SK; Feinstein LH; Heidmann,P. Healthy pets, healthy people. *Journal of the American Veterinary Medical Association*. 1999, v. 215(6): pp.335-338. Disponível em: <http://www.anthrozoology.org/healthy_pets_healthy_people>PMID: 10434969.

In: Dogs
Editors: K. M. Cohen and L. R. Diaz

ISBN: 978-1-62808-530-3
© 2013 Nova Science Publishers, Inc.

Chapter 5

ADVANCES IN THE CANINE COPROPARASITOLOGICAL EXAMINATION

Jancarlo Ferreira Gomes[1,2],
Celso Tetsuo Nagase Suzuki[2], *Alexandre Xavier Falcão*[2],
Sumie Hoshino Shimizu[3],
Willian Marinho Dourado Coelho[4],
Daniel Fontana Ferreira Cardia[4],
Sandra Valéria Inácio[4]
and Katia Denise Saraiva Bresciani[4]

[1]Departamento de Biologia Animal do Instituto de Biologia da
Universidade Estadual de Campinas (UNICAMP),
São Paulo, Brasil
[2]Departamento de Sistemas de Informação do Instituto de Computação da
Universidade Estadual de Campinas (UNICAMP),
São Paulo, Brasil
[3]Faculdade de Ciências Farmacêuticas da Universidade de São Paulo
(USP), São Paulo, Brasil
[4]Departamento de Apoio, Produção e Saúde Animal da Faculdade de
Medicina Veterinária de Araçatuba (FMVA) da Universidade Estadual
Paulista (UNESP), São Paulo, Brasil

ABSTRACT

In this review of advances in the parasitological examination of feces of humans and animals, are three important projects: two techniques that provide diagnostic improvements, and a project that stands out for its originality in the world. The first project is the *TF-Test* technique (Three Fecal Test), which is able to identify a large numbers of parasitic structures in fecal material from humans and animals (dogs and sheep). This technique showed high diagnostic efficacy, surpassing by more than 24% of sensitivity the association of conventional techniques and a commercial kit. The second project concerns the new *TF-Test Modified* technique, which, according to partial research results, demonstrates higher sensitivity than the *TF-Test* technique and other standard routine laboratory techniques. Finally, the third featured project is the automation of diagnosis of intestinal parasites by computerized image analysis. Preliminary studies have shown that this diagnostic system has high accuracy in terms of sensitivity, specificity, efficiency and concordance index κ, in the automatic detection of the 15 most prevalent parasitic species in Brazil. The results presented here demonstrate the diagnostic efficiency of *TF-Test* and *TF-Test Modified* parasitological techniques, and the future perspective of an automated diagnostic system unprecedented in the world.

1. INTRODUCTION

Canine and human intestinal parasites are highly prevalent worldwide, especially in tropical regions of the planet, where are underdeveloped and developing countries (Garcia, 2007; Chomel, 2008; Katagiri and Oliveira-Sequeira, 2008; WHO, 2010). In these regions, dogs are among the most preferred pets (Singh et al., 2004), and seem to have an important role in the transmission of more than 60 zoonotic infections, being included among these infections the ones caused by various genres of intestinal parasites such as *Giardia* spp., *Ancylostoma* spp., *Entamoeba* spp., *Cryptosporidium* spp., etc. (Katagiri and Oliveira-Sequeira, 2008; Rodie et al., 2008; Coelho et al., 2011). According literature reports, the estimated population of dogs in the world is more than 700 million, reaching about 10% of the human world population (Chomel, 2008; Arias, 2009). In clinical laboratories, several diagnostic tools are used to reveal the presence of intestinal parasites. However, the parasitological stool examination constitute the most commonly used exam to detect parasitic infection agents, together with the host clinical data, because of its simplicity, since it presents good sensitivity and allows to identify (direct result) the causative agent of the disease (Garcia, 2007; Katagiri and Oliveira-Sequeira, 2010).

Currently, in veterinary and human parasitology, the stool examination techniques are most often processed in a qualitative manner, by means of different laboratory principles for parasitic concentration and debris elimination, such as: centrifugal sedimentation, centrifugal flotation, spontaneous flotation, and spontaneous sedimentation (Garcia, 2007; Arias and Angel, 2009; Katagiri and Oliveira-Sequeira, 2010). In the field of veterinary medicine, especially for the diagnosis of canine feces, the most used techniques are supported in the qualitative principles of centrifugal flotation (Sheather, 1923; Faust et al., 1938), spontaneous flotation (Willis, 1921) and direct examination (Hendrix and Robinson, 2006). In turn, in human parasitology, there are several commercial kits and conventional techniques employed, which demonstrate the same qualitative principles previously mentioned for dogs, such as: centrifugal sedimentation (Ritchie, 1948; *TF-Test* - Gomes et al., 2004); centrifugal flotation (Faust et al., 1938); spontaneous flotation (Willis, 1921); spontaneous sedimentation (Lutz, 1919/Hoffman et al., 1934), and direct examination (Garcia, 2007). In this area, two principles are also featured in laboratory routine: A technique shown as quantitative for the detection of helminth eggs (Kato-Katz - Katz et al., 1970); and a principle related to the qualitative identification of larvae of intestinal nematodes, which is based on positive thermo-hydrotropism (Baermann, 1917; Moraes, 1948).

However, according to literature data and mainly due to changes in the epidemiological profile of parasites, recent studies have shown that conventional coproparasitological techniques and commercial kits used in the exam of animals and humans may leave to be desired in diagnostic efficacy, for demonstrating low or moderate diagnostic sensitivity (Gomes et al., 2004; Hendrix and Robinson, 2006; Katagiri and Oliveira-Sequeira, 2010; Brandellia et al., 2011; Carvalho et al., 2012).

Our main concern in this chapter was to present successful results of projects that come from scientific and technological innovation in the diagnostic field of Veterinary and Human Parasitology, especially through a literature review on recent advances in the parasitological examination of feces of dogs and humans. Two projects provide diagnostic improvements (Gomes et al., 2004; Falcão et al., 2010), and one project, although still under development and validation, stood out for its originality in the world (Falcão et al., 2008; Gomes et al., 2010; Suzuki et al., 2012).

2. ADVANCES IN THE COPROPARASITOLOGICAL EXAMINATION

A new coproparasitological technique was validated in 2004 (*TF-Test* - Three Fecal Test), initially for the diagnosis of human feces (Gomes et al., 2004). This technique, which contains in its operational protocol a kit (Figure 1), allows triple fecal sample collection on alternate days (every other day) and a single laboratory processing, with procedures of parasite concentration and debris elimination, through the centrifugal sedimentation principle. In studies conducted in different regions of Brazil, the aforementioned parasitological technique was able to provide moderate to high efficiency in diagnostic tests performed in humans, exceeding by more than 24% of sensitivity other conventional techniques and commercial kit (Gomes et al., 2004; Carvalho et al., 2012). In tests conducted in dogs, this technique showed lower diagnostic accuracy than when evaluated in humans, as reported by Gomes et al. (2006) and Katagiri and Oliveira-Sequeira (2010), although it excelled in the examination of feces of sheep (Lumina et al., 2006).

In 2010, Falcão and colleagues modified the conventional *TF-Test* technique, which was named *TF-Test Modified* (Falcão et al., 2010). Basically, two stages of concentration were added in this technique: spontaneous flotation and spontaneous sedimentation. In dogs, according to Coelho et al. (2013), this new technique has demonstrated high diagnostic efficacy. Similarly to humans, this technique showed relevant numbers in two studies, with high parasite concentration and elimination of most of the fecal debris (Rebolla et al., 2011; Carvalho et al., 2012) (Figure 2).

Finally, we present an unprecedented project in the world, which goes through the stages of final development and validation, with support from São Paulo Research Foundation - FAPESP (FAPESP, 2013). The objective of this project is to approach the automated diagnosis of intestinal parasites and, according to preliminary results of the study of 15 most prevalent parasitic species in Brazil, have shown high values for sensitivity (93.00%) and specificity (99.17%), with almost perfect kappa agreement (0.84) (Gomes et al., 2010; Suzuki et al., 2012; Suzuki et al., 2013).

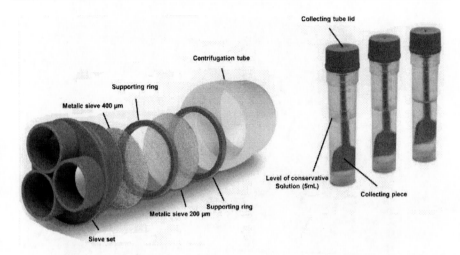

Figure 1. Parts of the TF-Test kit.

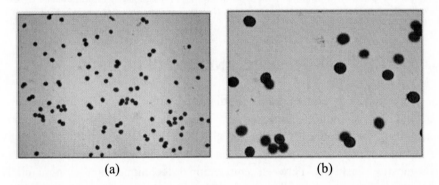

Figure 2. (a) Image field with large amount (over 95) of eggs of Hymenolepis diminuta obtained through the TF-Test Modified parasitological technique, with 4X microscope magnification; (b) Same slide, with 10X microscope magnification.

This automated diagnostic system, named by the research team as DAPI (in Portuguese, Automated Diagnosis of Intestinal Parasites), is composed of specific parasitological technique (*TF-Test Modified*), computer techniques for image analysis, computer, and custom equipment that consists of high-resolution digital camera, optical components and motorized stage (Figure 3).

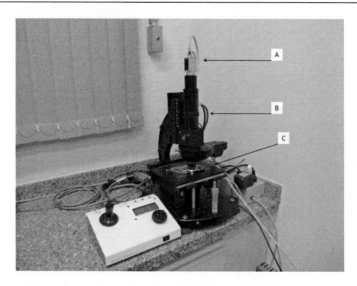

Figure 3. Custom equipment for the DAPI automated diagnostic system, consisting of: (A) high-resolution digital camera; (B) Optical components; and (C) Motorized stage.

3. RESULTS AND DISCUSSION

In canine or human parasitic diseases with high infection intensity, qualitative techniques employed in stool tests generally showed good diagnostic performance (Coelho et al., 2011). In this condition, a study (Gomes et al., 2004) performed on humans showed little variation in diagnostic sensitivity between conventional techniques and commercially available kits, such as between *TF-Test* (96%), Coprotest® (91%), StillQuick® (92%), Parasep® (87%), and Lutz/Hoffman (82%). However, under conditions of moderate and low intensity of parasitic infection, this same study showed favorable results for the technique with triple fecal sample collection and parasite concentration (*TF-Test*). Almost similar results were obtained by Carvalho et al. (2012), in a comparative study performed on humans between the coproparasitological techniques of *TF-Test*, Kato-Katz, Lutz/Hoffman, Willis and Baermann-Moraes.

The use of triple fecal collection, preferably on different days within a period not exceeding 10 days, significantly increases the diagnostic sensitivity of the parasitological examination of stools (Nazer et al., 1993; Hiatt and Markell, 1995; Cartwright, 1999; Gomes et al., 2004; Carvalho et al., 2012; Coelho et al., 2013). According to Gomes et al. (2004), the triple sample collection gave an increase in diagnostic sensitivity of about 24.41% (88.10% versus 63.69% of the second best technique) for the *TF-Test* technique, when compared with the combination of three conventional techniques (Lutz/Hoffman, Faust et al., and Rugai et al.) and a commercial kit (Coprotest). Coelho et al. (2013), applying the triple sample collection with the *TF-Test Modified* technique, showed an increase of 20.63% in diagnostic sensitivity in a study with feces of 106 dogs kept under control in a kennel. Moreover, in a study by Cartwright (1999), the triple fecal sample collection was able to increase the diagnostic sensitivity by 19.41%.

The difference that triple sample collection can provide in diagnostic sensitivity to the stool examination is evidenced in the study with dogs conducted by Katagiri and Oliveira-Sequeira (2010), wherein the authors did not collect samples on alternate days as recommended by the *TF-Test* technique protocol. Under normal conditions, this technique could certainly get more meaningful results, in accordance with recent study in dogs by Coelho et al. (2013).

Quantitative techniques can fail due to low infection intensity and due to characteristics of fecal elimination by the host, for example, quantity and consistency. In studies conducted with humans, Luis Rey (2008) and Kongs et al. (2001) reported low diagnostic efficacy of the Kato-Katz quantitative technique for estimating eggs per gram of feces of *Schistosoma mansoni*. These works are considered to be relevant to many quantification techniques in the field of Veterinary Parasitology, for example, Gordon & Whitlock Modified with feces of sheep (Ueno and Gonçalves, 1998), and McMaster and FLOTAC with feces of dogs (Cringoli et al., 2011), since, as for humans, the techniques may depend upon the characteristics of fecal elimination of the animal.

Another relevant issue, which can directly influence the diagnosis result, would be the choice of the technical principles for parasitic concentration and debris elimination. According to Garcia (2007) and Neves (2009), the correct procedure would be the microscopic reading of all fecal material collected and processed in laboratory. This does not occur in the use of conventional parasitological techniques and commercial kits, which, mainly for cost reduction and increased productivity, use only one parasitic concentration principle and a small amount of stool for microscopic analysis. These authors also reported that, in this situation, the examination should reveal only 30% to 50% of parasitic structures present in fecal material. Gomes et al. (2004) confirmed the comments of these authors in a study of 1102 individuals (humans) in four endemic regions for helminth infections. In this work, the combination of three conventional parasitological techniques (Lutz/Hoffman, Faust et al., and Rugai et al.) did not exceed 48.10% of diagnostic sensitivity. The moderate sensitivity presented by these techniques relate to the use of only one parasitic concentration technique, and use of a single fecal collection.

For example, the Lutz and/or Hoffman techniques, processed by spontaneous sedimentation, which use about 5g of feces in laboratory processing and about two drops (about 60µL) of fecal sediment for microscopic reading (Hoffman et al., 1934; Lutz, 1919), read less than 10% of the processed sediment, resulting in waste in microscopic reading and loss of diagnostic efficiency. Moreover, the Faust et al. and Sheather techniques, which use the centrifugal flotation principle, are based on the specific density of the highly saturated solutions of zinc sulphate and sucrose, respectively. The density of these solutions is close to 1.18g/ml, which causes the parasitic structures to float selectively. This, perhaps, is the great potential of this technical principle for detection of canine parasitic structures, which in most cases demonstrate specific densities lower than the density of the reagents mentioned above. In the studies by Katagiri and Oliveira-Sequeira (2010), and Coelho et al. (2013), they showed high diagnostic sensitivity for the Faust et al. technique. However, the centrifugal flotation technique would not be useful for the detection of parasites with densities greater than 1.18g/ml, for example, egg packets of *Dipylidium caninum*.

The use of more than one parasitic concentration principle certainly results in a gain of diagnostic sensitivity. Recently, with the use of the new *TF-Test* Modified technique, which takes three technical principles (centrifugal sedimentation, spontaneous flotation, and spontaneous sedimentation), Coelho et al. (2013), in the study of dog feces, and Carvalho et al. (2012) and Rebolla et al. (2012), in the study of human feces, demonstrated the superiority of this new technique when evaluated in comparison with Faust et al., Willis, Kato & Miura (direct exam), TF-Test *Conventional*, Rugai et al., and Kato-Katz (currently named Helm-Test) techniques.

Sophisticated methods such as immunological and molecular, currently under development, are highly sensitive but have specificity problems, mainly because their methods are based on indirect evidence (Garcia, 2007), and not on the direct visualization of the infectious agent. However, there is an expectation of developing low cost immunological or molecular methods, which establish the difference between an active infection and a past infection, as well as for use in parasitological control and therapeutic programs.

Although conventional techniques and commercial kits have brought innovations to the coproparasitological exam of feces of animals and humans, there is a clear need for a diagnosis that has high diagnostic efficiency, in order to ensure the treatment and control of transmission of intestinal parasites (Gomes et al., 2004, Hendrix and Robson, 2006; Lumina et al. 2,006; Garcia, 2007). An alternative presented to solve this problem would be through the use of automated diagnostics, employing computer in order to reduce diagnostic error.

Some techniques for automatic analysis of biological samples are known, however, being far from the reality of a clinical laboratory, according to public databases of patent information. For example, the patent US6005964 discloses a system for automatically detecting intestinal protozoa in the water samples, with the use of computer controlled microscope. However, this system only detects protozoa in material free of fecal debris, which facilitates the use of computational techniques for image analysis. Furthermore, this patent is not for the automatic detection of intestinal helminths. The patent WO03102210 describes techniques for identifying abnormalities in cells, using computer, microscope and digital camera. However, this patent only performs detection of cells present in blood and free of fecal debris. The same comment applies to the patent FR2572528, which describes a system designed to count the cells present in blood, and bacteria in water and milk, also free from fecal debris, using a microscope and computer. Moreover, the patent RU2123682 presents a technique for automatic parasitic detection, using dyestuff, fluorescent microscope, video camera and computer. This patent differs from the parasitological examination of the feces because it performs an indirect detection, through the use of reactions between antigens and antibodies for the true diagnosis. Finally, patents EP0774734 and US20060133657 present systems for visual analysis of generic objects in images, with the use of microscope, camera and computer. Therefore, they do not specifically apply to the automatic diagnosis of intestinal parasites.

According to the above, it is important to clarify that there is no automated diagnostic system, or semi-automated, in the world, that performs the coproparasitological examination of feces of animals and humans. However, this new type of diagnosis has been a research topic in Brazil (FAPESP, 2013). For this, an innovative diagnostic system, called DAPI, undergoes development and validation. According to the literature (Gomes et al., 2010; Suzuki et al., 2012; Suzuki et al., 2013), the preliminary results of evaluation of DAPI in the study of 15 most prevalent parasitic species in Brazil have shown high values for sensitivity (93.00%) and specificity (99.17%), with almost perfect kappa agreement (0.84). Considering the moderate sensitivity displayed by manual techniques, surely this new automated system will become gold standard in the areas of Human and Veterinary Parasitology. This system should provide uniqueness to the parasitological examination of feces of humans, and later on in animals, especially by providing high diagnostic sensitivity, automatic fecal smear reading, and releasing the test results in a short period of time and with images of detected parasites, never actually seen in the world.

REFERENCES

Arias, C; Angel, BJ. A wide diversity of zoonotic intestinal parasites infects urban and rural dogs in Neuquén, Patagonia, Argentina. *Vet Parasitol*, 167, 81-85, 2009.

Baermann, G. Eine einfache methode zur auffindung von Ankylostomum (nematoden). In: Larven in erdproben. *Neded Geneesk Laborat Weltever Feestbundel*, 41, 1917.

Brandellia, CLC; Cargnina, ST; Willers, DMC; Oliveira, KRP; Tasca, T. Comparison between spontaneous-sedimentation method and Paratest for the diagnosis of intestinal parasitic infections. *Transactions of the Royal Society of Tropical Medicine and Hygiene*, v.105, 604 - 606, 2011.

Cartwright, CP. Utility of multiple-stool-specimen ova and parasite examinations in a high prevalence setting. *Journal of Clinical Microbiology*, v.37, n.8, 2408 - 2411, 1999.

Carvalho, GLX; Moreira, LE; Pena, JL; Marinho, CC; Bahia, MT; Machado-Coelho, GLL. A comparative study of the *TF-Test*, Kato-Katz, Hoffman-Pons-Janer, Willis and Baermann-Moraes coprologic methods for the detection of human parasitosis. *Memórias do Instituto Oswaldo Cruz*, v.107, n.1, 80 - 84, 2012.

Carvalho, JB; Santos, BM; Gomes, JF; Suzuki, Hoshino-Shimizu, S; Falcão, AX. Avaliação de uma nova técnica (*TF-Test Modified*) destinada ao diagnóstico de parasitoses intestinais. *Cong Bras Intern Biomed*, pôster (grupo 6), São Paulo, 2012.

Chomel, BB. Control and prevention of emerging parasitic zoonoses. *Int. J. Parasitol.* 38, 1211-1217, 2008.

Coelho, MDC; Gomes, JF; Amarante, AFT; Bresciani, KDS; Lumina, G; Hoshino-Shimizu, S; Leme, PL; Falcão, AX. A new laboratorial method for the diagnosis of gastrointestinal parasites in dogs. *Rev Bras Parasitol Vet*, 22 (1), 1-5, 2013.

Coelho, WMD; Amarante, AFT; Apolinario, JC; Coelho, NMD; Lima, VM; Perri, SHV; Bresciani, KDS. Occurrence of Ancylostoma in dogs, cats and public places from Andradina city, São Paulo, State, Brazil. *Revista de Medicina Tropical de São Paulo*, v.53, n.4, 181-184, 2011.

Cringoli, G; Rinaldi, L; Maurelli, MP; Morgoglione, ME; Musella, V; Utzinger, J. *Ancylostoma caninum*: calibration and comparison of diagnostic accuracy of flotation in tube, McMaster and FLOTAC in faecal samples of dogs. *Experimental Parasitology*, 128, 32-37, 2011.

Falcão, AX; Gomes, JF; Hoshino-Shimizu, S; Suzuki, CTN. Method of Preparation of coproparasitological fecal sample and clarifier Composition. *Patent Cooperation Treaty*: PCT BR2010/000340), (WIPO http://www.wipo.int/pctdb), 2010.

Falcão, AX; Gomes, JF; Suzuki, CTN; Papa, JP; Hoshino-Shimizu, S; Dias, LCS. A System for Diagnosing Intestinal Parasites by Computerized Image Analysis. *Patent Cooperation Treaty*: WO2008/064442). (WIPO http://www.wipo.int/pctdb), 2008.

Faust, EC; D' Antoni, JS; Odon V. A critical study of clinical laboratory techniques for the diagnosis of protozoan cysts and helminth eggs en feces. I. Preliminary communication. *Am J Trop Med*, 18, 169-183, 1938.

Fundação de Amparo à Pesquisa do Estado de São Paulo - FAPESP. Automatização do diagnóstico de parasitos entéricos do homem por análise computadorizada de imagens. (http:://www.fapesp.br), proc.n. 2011/51467-0, São Paulo, 2013.

Garcia, LS. *Diagnostic Medical Parasitology*. Washington D.C., published by ASM, USA, 1. 202, 2007.

Gomes, JF; Hoshino-Shimizu, S; Dias, LCS; Araujo, AJSA; Castilho, VLP; Neves, FAMA. Evaluation of a novel kit (TF-Test) for the diagnosis of intestinal parasitic infections. *Journal of Clinical Laboratory Analysis* (USA), v.18, 132-138, 2004.

Gomes, JF; Lumina, G; Amarante, AFT; Hoshino-Shimizu, S; Leme, DP; Dias, LCS; Falcão, AX; Britto, LMP; Mazzochi, RE. Using the *TF-Test Kit* for the parasitological diagnosis in dogs (*Canis familiaris*). *Coleg. Bras. Parasitol. Vet.*, several - abstracts 001 a 043, 406, 2006.

Gomes, JF; Suzuki, CTN; Papa, JP; Hoshino-Shimizu, S; Falcão, AX. Toward automation of the diagnosis of enteroparasitosis via computational image analysis. Medimond s.r.l., Pianoro, v. M815L5, 169-174, 2010.

Hendrix, CM; Robinson, E. *Diagnostic Parasitology for Veterinary Technicians*. 3rd Edition, Mosby, USA, 2006.

Hiatt, RA; Markell, EK. How many stool examinations are necessary to detect pathogenic intestinal protozoa? *American Journal of Tropical Medicine and Hygiene*, v. 53, n.1, 36 - 39, 1995.

Hoffman, WA; Pons, JA; Janer, JL. The sedimentation-concentration method in schistosomiasis mansoni. Puerto Rico, *J Publ Hlth*, 9, 281-298, 1934.

Katagiri, S; Oliveira-Sequeira, TCG. Comparison of three concentration methods for the recovery of canine intestinal parasites from stool samples. *Exp Parasitol*, 126, 214-216, 2010.

Katagiri, S; Oliveira-Sequeira, TCG. Prevalence of Dog Intestinal Parasites and Risk Perception of Zoonotic Infection by Dog Owners in São Paulo State, Brazil. *Zoonoses Public Health*, 55, 406-413, 2008.

Katz, N; Coelho, PMZ; Pellegrino, J. Evaluation of Kato's quantitative method through the recovery of *Schistosoma mansoni* eggs added to human feces. *J Parasitol*, 56, 1030-1033, 1970.

Kong, A; Marks, G; Verlé, P; Van der Stuyft, P. The unreliability of the Kato-Katz technique limits its usefulness for evaluating *S. mansoni* infections. *Tropical Medicine and International Health*, v.6, n.3, 163 - 169, 2001.

Lumina, G; Bricarello, PA; Gomes, JF; Amarante, AFT. The evaluation of "*TF-Test*" Kit for diagnosis of gastrointestinal parasite infections in sheep. *Brazilian Journal of Veterinary Research and Animal Science*, v.43, 496-501, 2006.

Lutz, AO. *Schistosoma mansoni* e a schistosomose, segundo observações feitas no Brazil. *Mem Inst Oswaldo Cruz*, 11, 121-155, 1919.

Moraes, RG. Contribuição para o estudo do *Strongyloides stercoralis* e da estrongiloidíase no Brasil. *Ver Serv Saúde Pública*, RJ, 1, 507-524, 1948.

Nazer, H; Greer, W; Sonelly, K; Mohamed, AE; Yaish, H; Kagawalla, PR. The need for three stool specimes in routine laboratory examinations for intestinal parasites. *British Journal of Clinical Practice*, v.47, n.2, 76 - 78, 1993.

Neves, DP. *Parasitologia Dinâmica*. Atheneu, São Paulo, 3ª Edição, 592, 2009.

Public databases of patent information. INPI (Instituto Nacional de Propriedade Industrial – http://www.inpi.gov.br), USPTO (United States Patent and Trademark Office - http://www.uspto.gov), EPO (European Patent Office – http://www.epo.org), JPO (Japan Patent Office – http://www.jpo.go.jp), CIPO (Canadian Intellectual Property Office – http://www.cipo.ic.gc.ca), FREE PATENTS ONLINE IP RESEARCH & COMMUNITIES (Patent search – http://www.freepatentsonline.com), visited on 05/03/2013.

Rebolla, MF; Silva, EM; Gomes, JF; Franco, RMB. Avaliação parasitológica em instituições municipais urbanas de ensino infantil com ênfase em *Giardia duodenalis*. XXII Cong Bras Parasitol, pôster protozoologia (n. 42), São Paulo, 2011.

Rey, L. Parasitologia. Guanabara Koogan, Rio de Janeiro, editado pela editora Guanabara Koogan, 883, 2008.

Ritchie, LS. An ether sedimentation technique for routine stool examination. *Bull US Army Med Dept*, 8, 326, 1948.

Rodie, G; Stafford, P; Holland, C; Wolfe, A. Contamination of dog hair with eggs of *Toxocara canis*. *Vet. Parasitol*. 152, 85-93, 2008.

Sheather, AL. The detection of intestinal protozoa and mange parasites by a flotation technic. *J Comp Ther*, 36, 266-275, 1923.

Singh, LA; Das, SC; Baruah, I. Observations on the soil contamination with the zoonotic canine gastrointestinal parasites in selected rural areas of Tezpur, Assam, India. *J. Parasitol. Dis*. 28, 121-123, 2004.

Suzuki CTN; Gomes, JF; Falcão, AX; Hoshino-Shimizu, S; Papa, JP. Automated Diagnosis of Human Intestinal Parasites using Optical Microscopy Images. *IEEE International Symposium on Biomedical Imaging: From Nano to Macro*, 2013.

Suzuki, CTN; Gomes, JF; Falcão, AX; Papa, JP; Hoshino-Shimizu, S. Automatic Segmentation and Classification of Human Intestinal Parasites from Microscopy Images. *IEEE Transactions on Biomedical Engineering*, v. 60, 803-812, 2012.

Switzerland, World Health Organization (WHO). Working to overcome the global impact of neglected tropical diseases. First WHO report on neglected tropical diseases. Genebra, 2010.

Ueno, H; Gonçalves, PC. Manual para diagnóstico das helmintoses de ruminantes. 4. ed. Tokyo: Japan International Cooperation Agency, 143, 1998.

Willis, HH. A simple levitation method for the detection of hookworm ova. *Med J Australia*, 29, 375-376, 1921.

In: Dogs   ISBN: 978-1-62808-530-3
Editors: K. M. Cohen and L. R. Diaz   © 2013 Nova Science Publishers, Inc.

Chapter 6

THE MANAGEMENT AND DOMESTICATION OF DOGS

José Antonio Soares[*1;] *André Luiz Baptista Galvão*[2],
Amanda Leal de Vasconcellos[2], *Elzylene Léga Palazzo*[3],
Thais Rabelo dos Santos[2], *Rodrigo Rabelo dos Santos*[2]
and Katia Denise Saraiva Bresciani[2,4†]

[1]FKB, Fundação Karnig Bazarian, Faculdades Integradas de
Itapetininga, Itapetininga, São Paulo, Brasil
[2]UNESP, Universidade Estadual Paulista, Faculdade de Ciências
Agrárias e Veterinárias de Jaboticabal, Jaboticabal, São Paulo, Brasil.
[3]FAFRAM, Faculdade "Dr. Francisco Maeda", Ituverava,
São Paulo, Brasil
[4]UNESP, Universidade Estadual Paulista, Faculdade de Medicina
Veterinária de Araçatuba, São Paulo, Brasil.

ABSTRACT

Currently, we notice an increasingly close contact between humans
and animals. A significant portion of the world population demonstrates
affective needs or choose to live alone and prefer to have a pet as
company, such as a dog, which plays an important role in these

* prof.soares@uol.com.br.
† Email: bresciani@fmva.unesp.br.

conditions, by its ease domestication. With this approach, nowadays is extremely important that the Veterinary Clinics and Pet Shops offer to their customers, i.e., the owners of these animals, effective conditions, by training courses and responsible ownership, among others, that make possible the raise, coexistence, that is good for both, owner and pet. The Management dawns as a master tool in organizing, planning and controlling, in order that Clinics and Enterprises, in general, look at this more ethically and offer good business, as well as to its customers and consumers.

INTRODUCTION

The veterinarian graduates as a clinician, but is in the leading of their clinics and pet shops. The high failure rate happens due to a lack of administrative aspects knowledge. Knowledge of marketing strategies can determine the success of a clinic and pet shop.

The finding is that, within a company, clinic, hospital or pet shop, there are only costs and the revenues come from the clients. The success is directly related to the perception that customers - the selected customers - and the management of these clients, valuing and stimulating adoption, will determine the operating results, the success of the business (Pereira, 2010).

An interaction approach should be adopted, where the relationships client-clinic imply to live experiences. The clinic will provide an atmosphere that reflects and makes the owner realize that the care and "climate" expresses the importance given to the client-patient complex. The interaction is based in the mutual cooperation between the clinic/hospital staff - not just clinical - and the client (owner), to give the patient the best conditions for the maintenance of health and wellbeing. In practice we illustrate the interaction approach as follows:

- Planning Marketing should be focused on key customers, rather than just in the service;
- The integrated care crew should be trained and the manager should be concerned to coordinate all aspects of the customer relationship;
- Marketing function of the Clinic should be more closely integrated with the customer's purchasing function (Pereira, 2010).

According to the Ministry of Agriculture, Livestock and Supply the pets are the most responsible for the increase in the gain of the veterinary industry.

They have significantly increased among Brazilian families, and helped the national economy. According to the Association of Product Dealers, service providers and defense for the Use of Animals (Assofauna) there are 29.7 million dogs and 14 million cats that encourage industries to renew and expand its product mix to better serve this great demand increasingly rigorous. In 2006, the gains of the veterinarian cluster was US$ 1.165 billion. This number has gradually increased. For the association's president, Jose Maria Parra, growth could be higher if not for the high tax burden that adds and hinders people's access to the product. Still, what we see is a meaningful representation of this market in economy of the country. Who benefits from this growth are the pet shops. Today, there are about nine thousand stores in Brazil. However, this number is insufficient to serve customers. Those who already exist, are seeking to create marketing strategies to win the competition, attracting customers and keeping them captive.

THE MARKET

To SEBRAE (Brazilian Service of Support for Micro and Small Enterprises), the entrepreneur should seek to identify, in the place where he want to install the business a commercial site available for rent, with adequate size and location.

Should also check the existence of similar businesses nearby, such as agricultural products sellers, veterinary clinics and supermarkets, that sell related products.

Another important point is to investigate the types of services offered by those who are already active in this segment and the prices level. For this, it is enough to survey budgets in the different establishments in operation.

The deal will depend directly on the choice of the type of service being provided by the Pet Shop company. The consumer market consists of potential customers, which consist primarily of families that live near the property. Residential customers are the right public for the pet shop company, thus being determining factor in the choice of products and services offered in the shop. So, the potential clients profile is drawn.

For the preparation of the pet shop business plan, a survey of local residents sample should be made, to identify their needs for services and interest in this business. Regions near to condominiums, middle and high class neighborhoods, having customers with greater purchasing power, to purchase products and services, tend to be regions with the greatest potential for this market. From the knowledge of these clients needs, is always possible to offer products and services appropriately. In general, these customers will be satisfied if the products and services offered have satisfactory quality and price.

Indicators of potential demand: when deciding to open a pet shop, the entrepreneur must take into account what the market offers as space. Locations already supplied by several companies serving the market do not show potential for new companies in the sector. For the case of the Pet Shop it is needed to take into consideration that the pet is in many cases, treated as a family member. Nowadays people are not just looking for feed for their animals, but specific care and ideal foods. So, this kind of market tends to stand out more and more, offering various types of products and services, such as: accessories and toys, consultations and medicines, rations, small animals, bathing, grooming and hotel for the interested public, as well by the offered advantages (personalized service, monthly payment, delivery, debit/credit payment, special discounts, quality and variety in products and services).

RESPECT AND MAKE THAT YOUR ENTIRE TEAM RESPECTS THE CUSTOMER'S TIME

Customers have commitments, set times to work, to take the kids to school and other activities that are important to them. The time is a precious and perishable good; once lost is unrecoverable (Pereira, 2010).

Customers are the lifeblood of any veterinary clinic. The professional must be aware and believe that he can reactivate clients, otherwise, when he notice, many others will be lost. Each customer is an associate, a partner of the business. Thus the professional can not neglect or the clinic may become uneconomical (Pereira, 2010).

To avoid slow and steady losses, the veterinary surgeon should monitor their number of active clients (with reasonable services and products purchase frequency) and new customer every month, evaluating reasons and client reasons of absence. It is Important to analyze the frequency of purchase of major customers. It is also indicated to verify if the clinic is managing to attract new pet owners who seek high level of service, at what rate, if this is satisfactory and is being built a lasting relationship with many of these customers. Important to know whether the clinical communication with customers has been occurring with the planned intensity (Pereira, 2010).

The professional should not believe assumptions but ask his client. The survey is expected to generate a range of information. The number of questions should not be excessive, six to ten questions, both in the telephone as written survey or personal interview. The questions should be short, requesting review of single topic. As an example: How satisfied are you with the clinic treatment? The explanations of the veterinarian in attendance were satisfactory? Questions vague or too broad should be avoided (Pereira, 2010).

The greater the number of respondents, better knowledge about the customer's point of view. To obtain a high response rate, customers must be requested to respond and must demonstrate an interest in participating, ensuring quality, spontaneity and sincerity to responses (Pereira, 2010).

Donald L. Beaver Jr. is the founder of a fast growing company in Tipton, Pennsylvania, USA, with a suggestive name: New Pig Corporation. For him, one should prefer complaints to compliments, a complaint is someone telling you that you have not satisfied him yet (Whiteley, 1992).

What Beaver mean, is that a claim or complaint may be an opportunity for differentiation in relation to the competitors and this can help you figure out how to meet your customers needs and desires that they do not find in other companies.

Whiteley (1991), in his book entitled "The Customer Driven Company", emphasize that we are always measuring, i.e., seeking to quantify what customer needs or wants and throws the following assumption: know why you're measuring, let the customers inform you what final results to measure, ask constantly how is your performance - and your competitor's, follow the internal procedures that should produce the results that clients say they wish - as well as the final results and inform your staff everything you find. In short, eliminate the barriers that prevent your employees to serve customers efficiently and predictably.

REGAINING CUSTOMERS

Therefore, it is valid to identify inactive accounts or those who have a tendency to inactivity, as well as to separate registry of all customers which do not visit the clinic for more than a year. A survey should be developed with customers to identify and tabulate the critical reasons that led to inactivity, in the form of a questionnaire composed by three or four questions, at most. Another measure is to use the attendant time available to call customers and ask questions, updating records. She must be trained for the task. A better option is to hire a telemarketing company that performs this survey by phone (Pereira, 2011).

The importance of customer loyalty is strengthened by the destructive potential of unsatisfied customers who stop buying with the company. These customers, besides not being more loyal to the company, may hinder the development of new business and the achievement of new customers (Zenone, 2010).

When creating a plan and organizational philosophy that allows the company to manage relationships with customers, i.e., the implementation of CRM (Customer Relationship Management) goes beyond loyalty programs or *call centers*. It becomes a corporate strategy, a continued and integrated long-term effort between all organizational areas, However, it is important to understand that the *call center* is an important step in the operational view of CRM; it is from it that the company establishes contact with the customer, not just getting market information, but also enabling direct communication with diverse audiences that interest the company (Zenone, 2010).

In Figure 1, one can see that CRM is more than just a loyalty program, there are other programs such as prospecting, promotional sales, and others that are also part of a strategic relationship with the market.

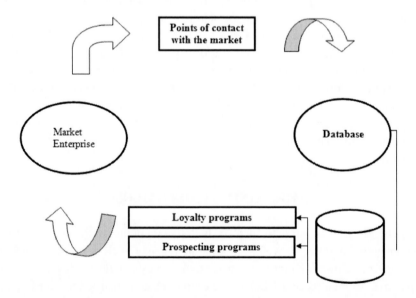

Figure 1. CRM Process and relationship strategies, adapted from Zenoni (2010, 63).

BILLING

Billing is the sum of all products/services sold by a company in a given period, and in this condition we call gross sales. Thus, there are many expenses and costs to be subtracted from revenues to get the salary and profit (Miracca, 2011a).

In the condition of self-employed we can admit that the monthly profit is equivalent to the salary, and in this case, it varies according to the number of attendances. Analyzing, salary is considered a fixed expense, which is present or not in billing, salary is not profit. Income is the remaining sum of gross sales, after deducting all expenses, costs, taxes, charges, including salary

To control the company finances, important tips are shown to the professional:

1) Understand that the professional and the company are different entities; that is why a legal person is created. The finances of both should be completely separate.
2) Maintain separate personal bank account from the company account.
3) Discovering what is the actual profit through careful control of all revenues and expenses (these controls can be achieved through spreadsheets or management software).
4) Determine a fixed amount for personal retreat. If this is not possible, consider the excess as "temporary loan" to be repaid as soon as possible to avoid decapitalize your company.
5) Adjust the amount of management fees to the profitability of your company. If that is not enough, you need to increase the company's profit.
6) Write down all withdrawals for personal expenses in the period, being the ideal to not withdraw anything until the end of the month and then remove only the amount of the management fees.
7) Remember that when you transfer your personal money to the company, you will be conducting a loan, the company will have to compensate later.
8) Create a reserve of capital to the company (capital, thirteenth salary and other labor expenses, equipment maintenance, new investments and extraordinary expenses) (Miracca, 2011a).

INCREASING REVENUES WITH VACCINATION, VERMIFUGE AND CASTRATION

Customers rarely request the application of the vaccine before the scheduled, leaving to do it days or months after the deadline. Some of them vaccinate elsewhere, because it is cheaper, convenient, or simply were seeking information from a competitor about some service (hotel, bathroom or even consultation) and take the opportunity. Thus, you lose a portion of those patients who were vaccinated in the previous year. Nevertheless, new patients also arrive and, then, are vaccinated animals unvaccinated in the previous year (Miracca, 2011b).

So, below are professional tips to increase the frequency of vaccinations:

1) Train your veterinary team to update and properly manage the information about the vaccine and vermifuge during the sessions.

2) Keep a track, especially for animals that frequent bathing and grooming clinic, reminding the customer, fifteen days before the expiration of the vaccine and vermifuge, and offering scheduled service on the same deadline.

3) Raise awareness, through contact by email, telephone or any other means with clients, advising them of the need for vaccination reinforcement, as well as the fortification of anthelmintic.

4) Educating the client about the advantages of keeping the vaccine and vermifuge of the pet up to date, emphasizing the benefits to the animal health and the economy to perform preventive medicine. This can be done during the service, through printed material, and other social networks.

5) When evaluating patients for vaccination, offer other services, treatment and cleaning of tartar, blood pressure control, etc.. In the situation of young, adult or elderly healthy dogs, clinically stable, that the owner does not want to place for mating, propose castration and emphasize the risk of developing pyometra in female dogs and canine prostatic hyperplasia for male dogs. For cats, castration reduces the risk of fights and escapes (Miracca, 2011b).

INVESTMENT IN EQUIPMENT

One of the basic principles of economics is that while resources are limited, the needs are endless. This means that it is never possible to do everything, and we will always have to choose between two or more things, and ideally make the choice that provides the best cost/benefit (Miracca, 2011c).

We should treat the acquisition of new equipment, such as investment and not as an expense, which means it will get essential financial return, enough to recover the invested amount, adjusted for inflation, in the shortest possible time, and still make a profit. An important issue to consider would be the opportunity cost - if instead of buying such equipment, the investment was made in another way, such as an increase in physical space of the clinic, could the profit be greater after two or three years?

In the purchase of equipment or new investments, the following topics should be considered:

1) Why do I need this equipment?
2) What will be the benefit for my customers?
3) This product may represent a new business opportunity?
4) How much can I charge and how I will offer this service or product?
5) Will I have additional costs, for example, when hiring a new employee to handle the equipment? What about costs and maintenance intervals?
6) How many customers will join the new service?
7) When purchasing imaging or diagnostic laboratory equipment, is the quality the same or higher than that offered by specialized centers?
8) When choosing various brands, should I select it by price or quality?
9) How long will it take to pay off the investment?
10) Are there any other investments that may be more profitable or are most important to my company? (Miracca, 2011c).

GENERAL CONSIDERATIONS

In the management of your business, the veterinarian, entrepreneur professional, must work constantly, be aware of the investment and, mainly, to the constant customer satisfaction.

The quest for customer loyalty should be a constant mission of the entire team, with the support of tools that help in practice and efficiency.

It is not enough to simply provide the expected product, it is needed to enchant them and, for it, various stimuli may be offered, from punctual promotions to a personalized service, where the customer and the consumer interact and feel happier and more satisfied.

The product or service may be the best on the market, but one wrong call, at the wrong time, can foul up a good negotiation and take the company to ostracism.

REFERENCES

Miracca, RB. 2011a. Faturamento, salário ou lucro? *Clínica Veterinária*, ano XVI, N° 92, 2011a, pp. 108-110, 1413571-X.

Miracca, RB. 2011b. Faturando mais com os mesmos clientes: o atraso da vacina. *Clínica Veterinária*, ano XVI, N° 93, 2011b, pp. 106-110, 1413571-X.

Miracca, RB. 2011c. Vamos às compras? *Clínica Veterinária*, ano XVI, N° 95, pp. 2011c, 108-110, 1413571-X.

Pereira, MS. 2010. Não é apenas a quantidade e sim a qualidade da comunicação que fará a diferença. *Clínica de Marketing – Nosso Clínico*, ano 13, N° 76, 2010, pp. 313-316, 1808-7191.

Pereira, MS. 2011.Clientes inativos. *Clínica de Marketing - Nosso Clínico*, ano 14, N° 84, 2011, pp. 345-348, 1808-7191.

In: Dogs
Editors: K. M. Cohen and L. R. Diaz
ISBN: 978-1-62808-530-3
© 2013 Nova Science Publishers, Inc.

Commentary

CAUGHT IN AN ACT OF CONVENIENCE: DISENTANGLING OUR THINKING ABOUT THE INFLUENCE OF OVARIOHYSTERECTOMY (SPAYING) ON HEALTHY LONGEVITY IN DOGS

David J. Waters[1,2,3]* *and Emily C. Chiang*[3]

[1]Department of Veterinary Clinical Sciences, and [2]Center on Aging and the Life Course, Purdue University,
[3]Center for Exceptional Longevity Studies, Gerald P. Murphy Cancer Foundation, West Lafayette, Indiana, US

INTRODUCTION

In our pets, we see the upside of domestication – an impressive quality of life and average lifespan that is the prized product of protection from infectious diseases, starvation, and predators. But the downside of domestication is that, like humans, highly protected canine populations experience the deleterious consequences of aging, including development of cancer and other age-related degenerative diseases [1]. When it comes to aging

---
* Corresponding author: David J. Waters DVM, PhD. Center for Exceptional Longevity Studies, Gerald P. Murphy Cancer Foundation, 3000 Kent Avenue Suite E2-100. West Lafayette, Indiana 47906. Phone: 765-775-1005; Fax: 765-775-1006; E-mail: waters@purdue.edu.

and cancer, pets and people are in the same boat. Now scientists and health professionals are beginning to place a high priority on gaining a better understanding of the aging process, finding the factors that can promote "healthspan" – living longer, healthier lives not just tacking on more years [2]. Increasing healthy longevity is becoming the goal for which humanity aspires.

To this end, in 2005, our research group at the Center for Exceptional Longevity Studies established The Exceptional Longevity Data Base, the first systematic study of the factors that favor highly successful aging in dogs. Instead of probing for interventions that might benefit geriatric pets, we committed ourselves to a novel approach to studying canine longevity: We focused on a *life course perspective on aging* – embracing the idea that early life events can profoundly influence adult health outcomes, including disease resistance and longevity. Our research sought to identify critical "windows" during the life course where superior early lifestyle choices and interventions could be applied to enable highly successful aging trajectories [3].

In 2008, this deeper sense of life course perspective led us to an important discovery: *No peer-reviewed manuscript in the veterinary literature had ever evaluated the association between canine longevity and the actual number of years of lifetime ovary exposure.* Instead, previous reports [4-5] exposed veterinarians to data on how long two groups of female dogs lived – "spayed" vs. "intact". "Intact" was the name given to bitches that were still sexually intact at the time of death. "Spayed" was the name given to bitches that lost their ovaries at some undetermined time during their lives. What became apparent to us was that ovariohysterectomy – the ovary-removing spaying procedure and elective act of endocrine organ excision widely advocated by DVMs in North America – had never been rigorously evaluated in terms of its impact on longevity.

In 2009, after carefully studying the association between the number of years of lifetime ovary exposure and highly successful aging in Rottweilers, *we discovered that keeping ovaries longer is associated with living longer* [6]. This link between ovaries and longevity was independent of lifetime investment in reproduction [7], as well as cause of death or familial longevity [6]. Our work pointed to a new line of thinking: *Ovaries are part of a system that promotes longevity.* This transformational way of thinking – seeing ovaries not just as reproductive units but as healthspan-promoting endocrine organs – is now supported by newer research on the longevity-extending effects of ovaries in women and mice [8-12].

Should Rottweilers be viewed as unreliable informants of the real relationship between ovaries and healthy longevity? Or could our nonconforming view derived from Rottweilers simply reflect that the method we used – analyzing the number of years of ovary exposure – is a more precise, health-relevant measure of interindividual differences in lifetime gonad exposure? It seemed prudent that we should probe this possibility. And in a follow-up study [13], we found that by using the common method of categorizing females as spayed or intact at the time of death (so-called dichotomous binning) – *ignoring the timing of spaying in each bitch* – a statistically significant relationship between number of years of ovary exposure and longevity could be distorted [13]. Our conclusion: The habit of veterinarians categorizing bitches as spayed or intact based upon gonadal status at the time of death is inadequate for representing important biological differences in lifetime ovary exposure, which can lead to misleading assumptions regarding the overall health consequences of ovariohysterectomy [13].

So how can we disentangle our thinking about ovaries and health amidst the new data supporting the potential longevity benefits of keeping ovaries? First, we need to broaden our thinking. It is time to expand our thinking of ovaries beyond reproduction, *seeing spaying as a physiological disturbance* capable of exerting system-wide effects [14]. Yet, early elective endocrinological disruption continues to be a widely recommended, "health-promoting" procedure for bitches in North America. If we can agree that the removal of endocrine organs (i.e., ovaries) can disturb normal physiology and physiological resilience in unforeseen ways, we encounter a fresh opportunity: *We might transform elective spaying from an act of convenience to a strategic disturbance – an intervention whose timing should be individualized to optimize each dog's chance of achieving healthy longevity.*

Second, by broadening our thinking, we can begin to change the dialogue. Instead of perpetuating the tiresome debate of whether spaying is "good" or "bad", finding the optimal window of ovary exposure for disease resistance and successful aging will become the prescient issue. Progress in science is measured not so much by the "facts" we generate, but by the new questions we ask [15]. And so it is that we must build the quality of our questions about spaying. When we ask "Is spaying good or bad?" or "Do spayed females live longer than intact females?", we pose the wrong questions. When we ask "What is the relationship between the *timing* of spaying and longevity?" or "What is the window of ovary exposure that will optimize healthy longevity?", we ask better ones. Categorizing bitches as spayed or intact without regard for the timing of spaying muddles thinking. We oversimplify biology and muddle information whenever we fall into the trap of either-or-ness [16]. The continued use of careless vocabulary can only hinder us as we seek to define the life-long, system-wide influences of ovaries on key cellular processes, health outcomes, and physiologic trade-offs.

It is difficult to predict just how well prepared the veterinary profession is to engage in this conversation, to skillfully weigh and consider new information on the biology of aging and healthy longevity. Veterinarians are not trained in the biology of aging as part of their DVM curriculum. So it might be expected that veterinarians would feel ill-prepared, reluctant to participate in any biogerontological debate. But if we have learned anything from the general semanticists – those experts who study how language shapes and limits perception – it is that we see the world through our categories [17]. And if we can express to veterinary students the idea that ovaries are more than just reproductive units, we arm them with a new and powerful categorizing scheme: Ovaries joining the ranks of thyroid glands, adrenal glands, and the insulin-producing pancreas as organs for which we generally advocate a retaining, a refraining from elective removal. This re-categorizing of ovaries as endocrine organs is a foundational step toward any serious re-evaluation of the lifelong health consequences of the act of elective ovary removal.

Finally, as our thinking and our conversations lead to new angles of vision, we will come to see that the ovary-longevity connection is an idea ripe for further inquiry. No longer will we be satisfied with review articles showcasing their bloated list of references, giving a false sense that "what we know" is extensive [18-19]. These offerings to practitioners and veterinary scientists have fallen far short of critically analyzing the relationship between lifetime gonad exposure and health outcomes, instead relying upon studies that categorize dogs as spayed or intact to "cover" the subject. Today, the subject calls for no more covering. *We need more uncovering, more discovering.* We must take action to advance the kinds of original investigative efforts that can provide a progressive framework for ongoing debate, future inquiry, and the pursuit of possibilities.

More than a quarter century ago, the Nobel Prize-winning immunologist Sir Peter Medawar wrote that all experimentation is criticism – the criticism that naturally arises from a dissatisfaction with prevailing beliefs [20]. Dissatisfaction beckons for disentangling. And as we grow to see the need for disentangling our thoughts about spaying, we prepare the ground for hastening a healthy reconsidering – not by imprisoning minds in an act of convenience, but rather by freeing them from their past wanderings. Paradoxically, this creative freedom can best be achieved through constraint: by staking the attention of veterinary scientists and practitioners in close proximity to the new ideas about ovaries. Here, we submit that continued investigation is sorely needed to more deeply understand the ovarian ecology that sustains healthier aging trajectories. Such inquiry is at the root of disentangling our thinking about how to effectively offset the downside of domestication – enhancing resilience and delaying the onset of age-related disease and disability by making more informed choices regarding the elective removal of endocrine organs.

REFERENCES

[1]   Waters, D. J. and Wildasin, K. (2006). Cancer clues from pet dogs. *Sci. Am.*, 295, 94-101.

[2]   Waters, D. J. (2011). Aging research 2011: exploring the pet dog paradigm. *ILAR J.,* 52, 97-105.

[3]   Waters, D. J. and Kariuki N. N. The biology of successful aging: watchful progress at biogerontology's known-unknown interface. In: Wilmoth, J. M. and Ferraro, K. F., editors. *Gerontology: Perspectives and Issues.* New York: Springer Publishing Company; 2013; 19-48.

[4]   Bronson, R. T. (1982). Variation in age at death of dogs of different sexes and breeds. *Am. J. Vet. Res.*, 43, 2057-2059.

[5]   Mitchell, A. R. (1999). Longevity of British breeds of dog and its relationships with sex, size, cardiovascular variables and disease. *Vet. Rec.,* 145, 625-629.

[6]   Waters, D. J., Kengeri, S. S., Clever, B., Booth, J. A., Maras, A. H., Schlittler, D. L., and Hayek, M. G. (2009). Exploring mechanisms of sex differences in longevity: lifetime ovary exposure and exceptional longevity in dogs. *Aging Cell,* 8, 752-755.

[7]   Kengeri, S. S., Maras, A. H., Suckow, C. L., Chiang, E. C., and Waters, D. J. (2013) Exceptional longevity in female Rottweiler dogs is not encumbered by investment in reproduction. *Age*, DOI: 10.1007/s11357-013-9529-8.

[8]   Parker, W. H., Broder, M. S., Chang, E., Feskanich, D., Farqhar, C., Liu, Z., Shoupe, D., Berek, J. S., Hankinson, S., and Manson, J. E. (2009). Ovarian conservation at the time of hysterectomy and long-term health outcomes in the Nurses' Health Study. *Obstet. Gynecol.*, 113, 1027-1037.

[9] Rocca, W. A., Gorssardt, B. R., de Andrade, M., Malkasian, G. D., and Melton, L. J. (2006). Survival patterns after oophorectomy in premenopausal women: a population-based cohort study. *Lancet Oncol.*, 7, 821-828.

[10] Rocca, W. A., Shuster, L. T., Gorssardt, B. R., Maraganore, D. M., Gostout, B. S., Geda, Y. E., and Melton, L. J. (2009). Long-term effects of bilateral oophorectomy on brain aging: unanswered questions from the Mayo Clinic Cohort Study of oophorectomy and aging. *Womens Health*, 5, 39-48.

[11] Cargill, S. L., Carey, J. R., Müller, H. G., and Anderson, G. (2003). Age of ovary determines remaining life expectancy in old ovariectomized mice. *Aging Cell*, 2, 185-190.

[12] Mason, J. B., Cargill, S. L., Anderson, G. B., and Carey, J. R. (2009). Transplantation of young ovaries to old mice increased life span in transplant recipients. *J. Gerontol. A Biol. Sci. Med. Sci.*, 64, 1207-1211.

[13] Waters, D. J., Kengeri, S. S., Maras, A. H., and Chiang, E. C. (2011). Probing the perils of dichotomous binning: how categorizing female dogs as spayed or intact can misinform our assumptions about the lifelong health consequences of ovariohysterectomy. *Theriogenology*, 76, 1496-1500.

[14] Waters, D. J. (2011). In search of a strategic disturbance: some thoughts on the timing of spaying. *Clin. Theriogenol.*, 3, 433-437.

[15] Pirie, N. W. Selecting facts and avoiding assumptions. In: Berthoff, A. E., editor. *Reclaiming the imagination*. New Jersey: Boynton/Cook Publishers Inc.; 1984; 203-211.

[16] Waters, D. J. (2012). The paradox of tethering: key to unleashing creative excellence in the research-education space. *Informing Sci.*, 15, 229-245.

[17] Johnson, W. *People in Quandaries*. New York: Harper and Brothers; 1946.

[18] Root Kustritz, M. V. (2007). Determining the optimal age for gonadectomy of dogs and cats. *J. Am. Vet. Med. Assoc.*, 231, 1665-1675.

[19] Root Kustritz, M. V. (2012). Effects of surgical sterilization on canine and feline health and on society. *Reprod. Domest. Anim.*, Suppl. 4, 214-222.

[20] Medawar, P. B. *Advice to a Young Scientist*. New York: Basic Books; 1979.

INDEX

E

F

J

K

L

M

N